]
of the OLD WEST

True Tales of the Old West

by

Charles L. Convis

Watercolor Cover by Mary Anne Convis

PIONEER PRESS, CARSON CITY, NEVADA

Library of Congress Catalog Card Number: 96-68502

ISBN 1-892156-13-X

Printed by
KNI, Incorporated
Anaheim, California

CONTENTS

ILLUSTRATIONS

PASSION IN DIPLOMACY

He was one of four men who made the most incredible journey into the unknown in North American history. Later he led the way on another journey that opened the southwest for settlement. He was a Muslim but he counted beads, muttered Hail Marys, and danced to the pagan gods of the Indians; so he was a blasphemer and a hypocrite. He drank a lot, he bedded Indian women at will, and he told wild, untrue tales around campfires; so he was immoral, a drunk, and a liar. But his courage, his tact, his diplomacy made him important in Old West history. This dark slave, Estevanico, was one of the bravest, most accomplished explorers of the New World.

He grew up in Morocco, probably born about 1500. He, his owner, and two other Spaniards were the only survivors of a 300-man expedition whose task in 1528 was to build two permanent settlements in Florida. After they failed to find treasure in Florida, they made their way to Texas in homemade boats, were enslaved by American Indians, and watched as their comrades died off, one by one. In 1534 the four survivors reunited to travel west and south, hoping to find the City of Mexico.

Estevanico did not act as a slave on their epic, two-year journey, the most daring and difficult in New World history. The account of the journey, published back in Spain, said he was as brave as a lion and as cunning as a panther. His unschooled mind was shrewd and quick, and he learned new languages faster than the others.

The natives thought the four travelers were Children of the Sun, and treated them as gods. Estevanico, a lusty man, had the most beautiful of the women assigned to him. And he showed his appreciation for their charms and skills. His three companions thought the role of gods should be played discretely and quietly; too much familiarity and sociability would lower them in the eyes of the Indians. Estevanico disagreed, and sometimes Cabaza de Vaca would scold him for brashness when spouses complained.

But the women never complained, and the dark man's warm amiability and eagerness to enter into native life as he found it produced information about the country, its resources, and the next tribes on ahead that — denied to his aloof companions — contributed greatly to their survival. How many progeny he left in his wake of passionate diplomacy is as unknown as it would be with York, the black slave with Lewis and Clark, two and a half centuries later.

In every Indian village they passed, the natives lavished gifts on the four travelers. Somewhere along the Colorado River in Texas they received gourds, bored with holes and containing pebbles. The gourds were used as rattles in medicine dances to heal the sick. The Indians did not know where the gourds came from; the river had washed them down from the west in a spring flood. Estevanico kept one of the gourds, as he was impressed with the beauty of its unusual feather and bead decorations. Throughout their journey, which ended in the City of Mexico on July 24, 1536, the travelers were told of fabulous cities of gold, silver, and jewels just ahead or not far to the north, but they never found any. Rumors of the cities had persisted in New Spain, and when the four survivors appeared from the north, the new viceroy, Antonio de Mendoza, resolved to send an expedition in search of the great riches.

Cabeza de Vaca returned to Spain, and his two white companions married wealthy widows and settled down to family life, each fathering eleven children. This left Estevanico who, upon reaching civilization, of course reverted to the status of slave. When the viceroy asked him to guide the expedition of Fray Marcos de Niza to the north, he gladly accepted. Now he could return to the free life he had known before! He had kept with him the highly decorated gourd he had received from the Indians in Texas. It was his favorite treasure.

The new, young governor of Nueva Galicia, Francisco de Coronado, helped plan the expedition. No soldiers would accompany it. It had a second purpose, equally important to the search for wealth. The northwest territories had been raided by Spanish slavers, and the Indians were to be assured by the expedition that the new viceroy would protect them and champion their claims for restitution. This second purpose would require diplomacy, and the viceroy would have liked to give the orders directly to Estevanico, but he trusted that Fray Marcos. — one of two Franciscan friars to go — would pass them on.

The viceroy told Fray Marcos: "You shall make clear to the Indians that I am sending you in the name of his Majesty to tell them that the Spaniards will treat them well, to let them know that he regrets the abuses and harm they have suffered, and that from now on they shall be well treated and those who mistreat them shall be punished."

This language from the account of the earlier Cabeza de Vaca journey suggests that the viceroy felt assured that Estevanico would transmit the message willingly: "We ordered them to come down from the mountains in confidence and peace, inhabit the whole country and construct their houses, including one for God. When Christians came among them they should go out to receive them with crosses in their hands, without bows or arms. The Christians would do them no injury, but be their friends."

The viceroy did tell Estevanico that he should obey Fray Marcos' orders as if they came from the viceroy, himself. He also told Fray Marcos to always learn before entering an area whether the natives in it were at war or peace with their neighbors. He did not want an unfortunate incident, provoked by a people already at war, to force a deviation from his program of kindness.

After passing the winter at Culiacan, the expedition left Coronado and his soldiers behind, setting out March 7, 1539, with Estevanico in the lead. Although he had been urged by the viceroy to obey Fray Marcos, he apparently had no intention of doing that. Once they were in the wilderness with the soldiers far behind, the friars would be dependent on him for survival. Overshadowed by Cabeza de Vaca on the earlier journey, now he would be a god in his own right. He dressed himself in bright feathers with bells tinkling at his ankles. A crown of plumes added to his natural height, and strings of coral, presented by admiring Indians along the way, swung from his chest. A growing retinue of admiring Indian girls — a regular harem — followed the colorful, brown-skinned man, his white teeth gleaming, as he strode forth in a regal manner. The friars heartily disapproved.

Remembering the viceroy's instructions about determining if the natives were at peace or war, Fray Marcos decided to let Estevanico scout on ahead. He was to explore the country and send back messages of what he found by crosses. Moderate news would result in a cross, five or six feet high. Important news would mean a cross twice that size. And a very large cross would mean he was on the trail of something better than anything in all of New Spain. The friars were probably glad to see Estevanico with his Indian servant and three hundred admiring followers, many of them women, take the lead. It made life simpler for the heralds of Christianity who had more modest ideas about relations with strangers.

Four days after Estevanico left, Indian messengers arrived with the first cross. They relayed the information that just thirty days' travel ahead of Estevanico was the first city, one called Cibola. Estevanico added his understanding that six more large cities lay beyond the first. Of course, we only know this from what Fray Marcos wrote. If he lied — which he certainly did later — or if he misunderstood, or if Estevanico lied or misunderstood, or if the Indians lied are things we will ever know.

Estevanico kept traveling ahead, sending crosses back, and never letting Fray Marcos catch up. Later the friar would charge that Estevanico wanted the sole glory and honor of discovering the Cities of Cibolo. Yet, each time Fray Marcos came to Indians whom Estevanico had reached, he was welcomed and treated well. The former slave was performing admirably as an advance man.

At the valley of the Rio Sonora, about 125 miles south of present Arizona, Estevanico said he would wait up. But he didn't. Fray Marcos wrote that he kept receiving messages and crosses, and he kept following, trusting they would reach the legendary cities. But later-learned geography revealed his blatant lies. The friar averaged less than twenty miles a day. Yet he said he made a side trip from the Valley of the Sonora, over four hundred miles, in five days. He also said he reached the first city of Cibola.

As they approached that city, Fray Marcos wrote, they heard from wounded Indians in the advance entourage that Estevanico had been killed. Before reaching Cibola, Estevanico sent his sacred gourd on ahead to assure the occupants he came in peace. When the chief saw the decorations on the gourd, he flung it down and shouted, "I know these people, for these cascabels (decorations) are not in our fashion. If they do not turn back, no man of them will remain alive."

Estevanico, confident in his customary diplomatic approach, came on and was killed with several of his followers.

Fray Marcos also described the first City of Cibola. He said it was on a plain at the base of a round hill, and it was bigger than the City of Mexico. He wanted to examine it more closely but realized if he was killed he would be unable to bring back the wonderful news to Mexico. He retreated.

In fact, Fray Marcos never got north of the Gila River. His description of the first City of Cibola may have come from Indians he talked to, or it may have come from the Coronado expedition which saw it the next year. He was a member of that expedition. But he didn't even record the description properly. It was on a round hill overlooking a plain, not on a plain at the base of a hill, as he wrote. Eventually he was branded, both in Spain and in New Spain as "lying monk."

Estevanico was killed at the Zuni pueblo of Hawikuh, about a dozen miles southwest of present Zuni, New Mexico. His delusional reign as a god ended at a dusty village, a tiny fraction of the size of the City of Mexico. If he had not accepted the strangely decorated gourd in Texas and presented it to the Zuni chief in New Mexico, one might wonder how much longer his passionate diplomacy would have worked on opening the southwest to settlement.

Suggested reading: John U. Terrell, *Estevanico the Black* (Los Angeles: Westernlore Press, 1968).

ZALDIVAR'S TATTOOED WOMAN

Early explorers knew little about the size of North America. Columbus, who had not yet reached the continent, thought he was in the East Indies, halfway around the earth. Two hundred years later, French explorers in the north thought the Pacific Ocean was only a few days' journey west of the Great Lakes.

But as early as the 1540s, the Spanish had a good idea of the continent's size. They owed it to an unnamed Indian slave woman.

Coronado's exploration in 1540-1542 extended from New Spain to present Kansas with a side trip to the lower Colorado River between California and Arizona. At the same time, De Soto explored from Florida to East Texas. The Indian woman provided the information that tied the two explorations together.

Almost three centuries later Sacajawea, another Indian slave woman far to the north, would render priceless service to Lewis and Clark. But the tattooed woman slave of Captain Juan de Zaldivar should also be honored in the annals of American exploration.

Zaldivar, one of Coronado's captains, bought the woman in Tigeux in 1540. Tigeux was a group of twelve Anasazi villages near the Rio Grande, north of where Albuquerque would be founded a century and a half later. The woman was originally from east Texas, probably a Wichita from her tattoos, although possibly a Caddo. We have no idea who captured her or how many times she had been traded before she reached the Anasazi villages.

After wintering at Tigeux, Coronado moved east in 1541, seeking Quivira. A Pawnee trader, living with the Anasazis, lied to Coronado and told him Quivira was a city of gold in the northeast. The expedition reached the staked plains of West Texas in May.

Shortages of food and water and the difficulty of travel led Coronado to divide his forces at the Palo Duro Canyon, where the Prairie Dog Town Fork of the Red River had carved down a thousand feet. Coronado took thirty-six men, including Zaldivar, and continued his search. The remaining party, under Don Tristan de Arellano, stayed in the canyon area for two weeks, hunting buffalo and drying meat for the return trip to Tigeux. During the hunt, with Zaldivar gone, the tattooed woman escaped.

She recognized the country, having passed through it earlier. She traveled southeast to the Salt Fork of the Brazos River. She followed that fork and the main Brazos to about present Navasota, Texas, where the river turns south. At that place she continued east until she reached her people, somewhere north of present Houston. She had traveled nine days.

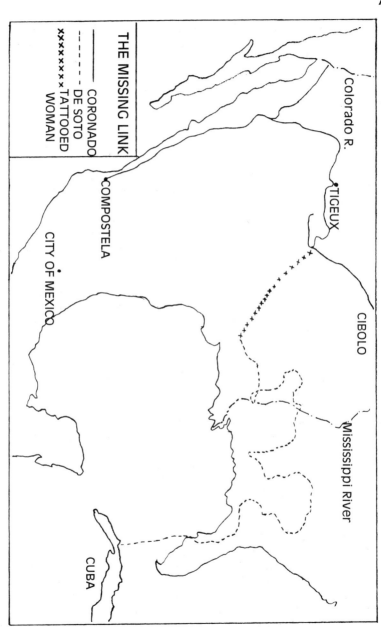

THE MISSING LINK

———— CORONADO
- - - - DE SOTO
×××××××× TATTOOED
           WOMAN

Colorado R.

•TIGEUX

CIBOLO

Mississippi River

CUBA

•COMPOSTELA

CITY OF MEXICO•

While Coronado was learning that the Indian villages on the Arkansas River in Kansas were not made of gold, De Soto was about three hundred fifty miles downstream on the same river, in present Arkansas. Each knew the other was a rival explorer; they had no idea how close they were to each other.

De Soto led his men back to the Mississippi, only to die and be buried in the huge river he had discovered. Luis Moscosa de Alvarado took command of the expedition, which decided to march overland to Mexico. They had no idea how far away it was.

When they reached East Texas in September 1542, Alvarado's expedition met the tattooed slave woman, who had escaped from Coronado's expedition the year before. She remembered traveling nine days, and she correctly repeated the names of some of Coronado's captains. The explorers knew she did not lie.

Alvarado's expedition turned back to the Mississippi, deciding the overland route would be too hazardous. The expedition completed the boats they had been building and sailed down the Mississippi and Gulf of Mexico to Tampico. From there they traveled overland to the City of Mexico.

The tattooed woman traveled about five hundred miles in nine days to reach her own people. She probably had to watch for enemies — Lipan Apaches and Tonkawas — during her incredible flight.

When Spanish map makers back in Spain got reports from the two expeditions, describing their travels, and added in the information which Alvarado obtained from the escaped slave woman, they knew how wide North America was.

De Soto had started in Florida and traveled north, west and south. Coronado had started in Compostela, a short distance from the City of Mexico and traveled north, west and east. They did not know how close they came to each other. But the map makers did after Zaldivar's tattooed slave woman provided the missing link.

Suggested reading: Herbert E. Bolton, *Coronado: Knight of Pueblos and Plains* (New York: McGraw-Hill, 1959).

KINGS OF THE WILD FRONTIER

The Chamuscado-Rodríguez expedition from New Spain into the wild lands of the north (New Mexico) in 1581 only had a dozen men. Too small to think of discovering great riches, the nine soldiers and three Franciscan friars settled for finding food to keep themselves going and heathen converts to keep the church growing. Also too small to think of conquering vast territory, they settled for staying alive while surrounded by thousands of hostiles. What they lacked in firepower they made up in courage, cunning, and bravado that would become the hallmark of later trappers, scouts, and frontier heroes in the best tradition of Davy Crockett.

They ran short of food when they reached a pueblo they called Piedra Aita and decided it was time to teach the natives about providing sustenance. The natives responded to their signs and sounds as if they were deaf. The expedition leader cautioned against force. Although the natives thought the invaders were Children of the Sun and therefore immortal, it was obvious that three thousand men would respond to a call to arms. Some caution seemed prudent, but the leader was ill and finally told his men they could do what they wanted as long as they didn't incite a revolt.

The natives proved adamant, so the leader rose from his sick bed and led seven mounted soldiers into the pueblo, ready for war. Apparently one stayed behind to guard the friars. A few soldiers fired their harquebuses, which roared and spat fire, and the natives quickly decided that each house would contribute a half peck of ground corn flour.

As they went on to other pueblos in the province, the soldiers were offered exactly the same amount of flour, no more, no less. Sometimes it was more than the soldiers could carry, so they graciously returned what was left over. The natives also contributed turkeys, of which they had many and which they did not value highly.

One friar, Fray Juan de Santa María, decided he wanted to return to the land of the Christians. The other eleven, not convinced by the natives' profession of friendship while turning over their food, tried in vain to change his mind. They said it would jeopardize their security and probably result in his being killed. He insisted, and the others were proved right. The natives thought the friar was going to get more Christians who might put them out of their homes. The soldiers learned after returning from a buffalo hunt that the friar had been killed after two or three days on his journey to the south.

The soldiers acted like they didn't understand what the natives were saying about killing the friar. Having discovered their visitors were no longer immortal, the natives were obviously preparing to kill the rest when the

Spaniards wisely decided to move on to another pueblo.

But at that pueblo, Indians visiting from a third pueblo, which the soldiers called Malagón, killed three of their horses. Determined that the crime should not go unpunished, the leader and five soldiers rode to Malagón to make their arrests and intimidate the natives.

Malagón had eighty houses, each three or four stories high. It had a plaza and streets and over a thousand residents. The angry soldiers, riding up in fighting order, could see the Indians keeping watch on their housetops. The soldiers demanded to know who had killed their horses. The Indians replied, "What horses?"

The soldiers fired their harquebuses, although being careful to not aim at any Indians. The frightened Indians ran inside their houses and began throwing out dead turkeys to placate their attackers.

About thirty Indians could still be seen on the roofs. To let them know they would not be placated, the soldiers demanded they be given the horses or the culprits who had killed them. The Indians insisted they had not killed any horses, asked the soldiers to not be angry, and declared that they wanted to be friends.

Three of the soldiers dismounted, while the other three stood guard, and started going through houses until they found the horseflesh. They went back to their horses, fired their harquebuses, and again demanded the culprits, who they said they would kill. If they didn't turn the culprits over, the soldiers would kill everybody. Now the Indians claimed the horses had been killed by people from another pueblo.

So the soldiers attacked! They captured two Indians who tried to run away. Before they took their prisoners back to the other pueblo, they decided to set fire to Malagón. One had a bit of hay lighted and ready to throw when they reconsidered and decided the whole pueblo shouldn't die for the crime of about a half dozen.

They decided to behead their two prisoners. As they set up a chopping block in the plaza and got a machete ready, they could see a thousand Indians solemnly watching the proceedings. The friars had already decided they would remain in that pueblo when the soldiers moved on. So someone with brains to match his courage suggested they should let the friars "rescue" the Indians. After a "tussle" the friars made the rescue, and were treated as saviors by the pueblo.

The next day a delegation from Malagón arrived to say they would kill no more horses. The Friars stayed in the pueblo and the soldiers rode on, apparently remaining in the area about as long as they wanted.

Suggested reading: Donald A. Barclay et al, *Into the Wilderness Dream* (Salt Lake City: University of Utah Press, 1994).

HERE BEFORE CHRIST

Seventeen-year-old Pierre Esprit Radisson was returning home to Three Rivers on the St. Lawrence River in New France when he came upon the bodies of two friends. They had all left the stockaded fort together that spring morning in 1652 to hunt. When Radisson wanted to move on to a dangerous looking place, his companions had held back, warning about the Mohawks, who tolerated the English and hated the French. As he looked on in horror at his two companions, stripped, scalped, their bodies hacked and torn, their Mohawk killers captured him.

His courage in resisting capture impressed the Indians, and they allowed him to run their gauntlet unharmed. A captive Huron woman in the lodge of a leading chief persuaded him to adopt the boy into their family. He was given a gun and allowed to hunt with the Indians.

While hunting with three warriors, they came upon an Algonquin slave of the tribe, who suggested that he and Radisson kill the warriors in their sleep. The French boy refused, but during the night the slave woke Radisson, put a hatchet in his hands, pushed him toward a sleeping Mohawk, and killed the other two warriors. Radisson struck at the Indian as he woke, killing him. The Algonquin scalped all three, and he and Radisson escaped in a canoe.

Fourteen days later they were within sight of Three Rivers when the Mohawks attacked. They threw out the scalps, which floated until the Mohawks fished them out of the water. The Indians killed the Algonquin, sunk the canoe, and captured Radisson again. Heading back to their own country, the Mohawks were joined by other Iroquois, who had with them twenty prisoners, two French, one white woman, and seventeen Hurons.

The prisoners were half dead when the Indians reached their village. Radisson was again taken by his adopted family, but this time he had killed a tribesman, and the penalty was two more days of fiendish torture. He survived, but the injuries resulting from being thrown into a fire kept him from walking for a month.

Accepted again as a warrior, Radisson went on raids with his Indian brothers. During a visit to the Dutch trading post at Fort Orange, at present Albany, New York, a soldier who spoke French saw the white skin under his war paint and offered to ransom him. Radisson refused, but two weeks later changed his mind. He walked away from his village and then broke into a run. After running all day and night he reached a Dutch village late the next afternoon. The Dutch kept him concealed from his pursuers until they got him to Manhattan. He sailed from there to Amsterdam and made his way to France.

Radisson was back in Three Rivers in spring, 1654. He learned that the outraged citizens of Three Rivers had burned a Mohawk chief to death. The Mohawks had then killed the governor and a company of Frenchmen, including Radisson's brother-in-law. His widowed sister had remarried Médard Chouart Groseillers and from this came one of the outstanding exploring partnerships in the history of North America. Radisson and Groseillers began exploring into the Great Lakes country together.

Traveling alone, they reached the Mississippi River in 1659, the first white men to see the upper reaches of that river. They may have even reached the Missouri, as they described meeting Indians like the Mandans. They also penetrated the unknown wilderness to the north as far as Hudson Bay.

In 1661, the partners asked the governor for permission to extend their operations to Hudson Bay. The governor demanded half the profits from their fur trade. They refused, and the governor ordered them to not leave Three Rivers. They left anyway.

They returned to Montreal in 1663 with a retinue of seven hundred Indians in three hundred sixty canoes, filled with the greatest catch of furs the French had ever seen. The governor fined the explorers ten thousand pounds sterling, which absorbed all the profit. The explorers went to France to appeal.

They told King Louis XIV that a vast empire in North America, filled with untold riches in beaver and other skins, lay at his feet. He turned down their appeal, and that is probably why more people in North America today speak English than French.

In 1665, after trying unsuccessfully to get ships for their proposed venture in Boston, Radisson and Groseillers set out for England for an audience with King Charles II. He was seeking a water passage across the northwest part of North America. High drama still followed the explorers.

They were captured by the Dutch on their way to England. They reached London just after the great plague. The English King authorized an expedition that would allow the explorers to trade in Hudson Bay while their English ships searched for a northwest passage to the Pacific Ocean. Then the Dutch tried in vain to persuade them to desert in favor of Holland. Temporarily set back by the war between England and Holland, they finally sailed away in 1668, Radisson in the *Eaglet* and Groseillers in the *Nonsuch*. They ran into an Atlantic gale which dismasted the *Eaglet*. It returned to England, and the *Nonsuch* sailed through Hudson Strait and Hudson Bay, reaching the south end of James Bay in September. There, very near the point that Radisson had reached five years before while traveling overland, Groseillers built Fort King Charles on the Nemiscau River.

In June the *Nonsuch* returned to England, loaded with furs. Leading London merchants had supplied the expedition in return for shares in the enterprise. The success of Groseillers' trading ensured the granting of a charter. It was issued on May 2, 1670 to the Company of Adventurers of England Trading into Hudson's Bay. Most of the shareholders were supporters of King Charles II, having helped him to the throne. The King's cousin, Prince Rupert, the Duke of York, headed the list, and the company's territory was often called Rupert's Land. Sir John Kirke, whose daughter had become Radisson's wife while he waited in London, was also a shareholder. The initials of the company and its long age resulted in it often being called Here Before Christ.

The company's territory included all the land drained by rivers flowing into Hudson Bay and Hudson strait. It included about half of present Canada, part of Minnesota, and a substantial part of North Dakota. The annual fee for trading in a territory larger than Russia was to be two elks and two black beavers, paid whenever the king or his heirs happened to visit the region.

In 1674 Radisson was surprised to learn that the company merely considered him and his brother-in-law as employees; they were not partners. Disappointed, Radisson accepted a French offer to return to that country. His English wife refused to move or change her loyalty. Soon, Radisson found himself at the center of a three-way international controversy between France, England, and the company. Officially the English tried to punish Radisson for selling out to France; unofficially, they tried to bribe him to return. France pardoned Radisson for his secession to England, paid his debts, and gave him a position in its navy.

All either country wanted, as they struggled to improve their positions in North America, was for Radisson to keep quiet, something the intrepid explorer was unable to do. Groselliers did retire to his family at Three Rivers. France then offered Radisson command of a man-of-war, but all he wanted was a commission to return to New France and the fur trade. The problem with this was his English wife. Her family had claims of forty thousand pounds against England, arising from the capture of Quebec. France would not give a commission to a man married to its enemy.

In 1681 Radisson returned to New France on his own and persuaded the governor to look the other way as he and Groseillers led another expedition to Hudson Bay. Groseillers' son, Jean Baptiste, went along. Despite an attack by a pirate ship (either Dutch or Spanish), straits filled with drifting ice, and mutinous crews, they reached Hayes River on the west shore of Hudson Bay — present York Factory — before winter. Radisson had been to the same place with English ships ten years before. He was glad to be back in the wilderness where brave men have the opportunity to assert dominion over nature.

Leaving his partner to build a fort, which they called Fort Bourbon, Radisson and his nephew canoed three hundred miles up the river to near Lake Winnipeg. Radisson had first visited this region twenty years before, and the Crees had looked forward to his return visits. Now they welcomed their old friend with great joy. They exchanged presents, cast all their furs at his feet, and promised to deiver more in the spring. The man who had struggled against jealousy and intrigue in London and Paris, now was showered with great respect and affection in the wilderness.

But they found more than Indians there! A short distance to the northwest, the Nelson River ran substantially parallel to the Hayes, also discharging into Hudson Bay. Radisson learned that a ship containing poachers from Boston was building a fort on the Nelson, and a Hudson's Bay Company ship, approaching to build their winter quarters, had run aground just nine miles downstream. The Hudson's Bay Company ship was commanded by a Captain Gillam, who had long opposed Radisson, and the poachers' were led by Ben Gillam, his son. Neither Gillam knew that the other was in the vicinity, and Radisson knew that if the two English speaking expeditions got together they would drive Radisson and his people out.

In an incredible display of bluff, bravado, and imagination, Radisson kept the two English speaking groups apart and, although greatly outnumbered, he captured both. Radisson obtained the entire fur crop when the Crees came in that spring. He could have got more when they offered an additional two hundred beaver skins for the privilege of killing the captured English- speakers. Radisson declined their offer.

But Radisson and Groselleirs had gone north without a permit, and their furs were seized in Quebec. Groselleirs, a tired old man, left the fight up to Radisson and went home to his family. Radisson, fifty, had a wife and four children to support. In spite of bringing furs worth a half million dollars in modern money to New France during a four year period and another half million to England in the ten years that followed, he was a poor man.

In 1684, Radisson took a new oath of loyalty to England. Then the French king ordered him arrested. For the rest of his life, Radisson divided his time between Hudson Bay and London.

History has called Radisson a traitor and a turncoat. But he was always faithful to his family, to his vision, and to his purpose in exploring new lands. The double dealing of France and England during his career far exceeded anything he could be fairly accused of.

Suggested reading: A. C. Laut, *Pathfinders of the West* (New York: Grosset & Dunlap, 1907).

FRANCE INVADES NEW SPAIN

The French, contesting with the English for what is now Canada, were first to reach the headwaters of the Mississippi River. Louis Jolliet and Father Marquette explored the river as far south as the Arkansas River.

René Robert Cavelier, Sieur de La Salle, a former Jesuit who asked to be released from his vows because of "moral frailties", was interested in exploring a possible route to China. In 1678 he got permission from King Louis XIV to follow the river to its mouth, build forts, and open communication with New Spain.

In 1682 La Salle sailed down the river to its mouth. He claimed the area for France and named it Louisiana after the king. He returned to France, which was at war with Spain, and falsely represented that the mouth of the Mississippi was very close to the Rio Bravo (Rio Grande), which everyone knew ran through the northern settlements of New Spain. He persuaded his king to establish a colony at the river mouth he had discovered. It would serve as a base for the conquest of New Spain, and from there the conquest of a continent. La Salle suggested that a fort on the Gulf of Mexico could be the starting point for an attack on the provinces, rich in silver mines but "far from help and defended by a small number of people, too indolent and effeminate to endure the rigors of such a war."

He sailed from France in July, 1684, with four ships, a hundred soldiers, and about three hundred colonists. The colonists included craftsmen and about a dozen families to establish farms. They also included single women recruited from French bordellos and men dragged out of the gutters of Paris, who had spent most of their lives in prison.

La Salle's naval commander, Captain Beaujeu, could not find the mouth of the Mississippi. The expedition landed in early 1685 at what is now called Matagorda Bay in Texas. Over the protests of Beaujeu, who knew better but had been unable to find the river's mouth himself, La Salle insisted that the river discharging into the bay was the Mississippi. Actually it was Cavallo Pass at the north end of Matagorda Island, where the waters of Matagorda Bay and Espiritu Santo Bay discharge into the Gulf of Mexico.

La Salle knew soon after he was ashore that they had not found the mouth of the Mississippi. But one of their ships ran aground in landing, and most of their supplies, tools, equipment, and ammunition were lost. Earlier another ship, heavily laden with supplies, had been captured by Spanish pirates. Beaujeu was anxious to leave, as he and La Salle had argued from the beginning of the voyage. A small frigate was left behind, and Beaujeu's

36-gun man-of-war sailed away, abandoning La Salle and his colonists on a desolate and hostile coast.

Many of the craftsmen, having lost faith in La Salle and thinking the expedition doomed, also returned with the man-of-war. So about two hundred men and twenty or thirty women and children were left to take possession of a territory half as large as Europe, a territory containing rich mines of gold and silver which the Spanish had held for a century and a half.

As La Salle built Fort St. Louis at the head of what is now called Lavaca Bay, his innate personal qualities assumed their darker aspects. Never capable of humor, he became morose. Always moral and devout, he became prudish and obnoxiously pious. His tenacity had now become obstinacy. Always strict with men in his command, he now inflicted brutal punishments to keep the wastrels, drunks, and libertines in line. He took few into confidence; all others he distrusted.

The fort was built from materials salvaged from the wrecked vessel. Located on sand dunes, it was unsheltered from the wind. Brackish green water, foul and stinking, seeped in everywhere. The colonists, eating mostly oysters and snakes, were weakened by scurvy.

They soon learned that the native Karankawas were not only hostile but also cannibals. Not knowing where the nearest Spanish settlement was, La Salle drove his colonists to complete the fort as though he were driving slaves. They spread a sail cloth to shield twenty men, dying in agony from syphilis contracted earlier during a stop over at Santo Domingo. Ten soldiers died from food poisoning. A craftsman vanished while hunting, apparently killed and eaten by Karankawas. Another man was bitten by a rattlesnake. His leg was amputated and he died two days later. The priests persuaded La Salle to house the single women in a separate, guarded hut. Several men then asked permission to marry, and the ceremonies were performed.

By October a small fort had been completed, and La Salle, with fifty men, set out along the coast to search for the Mississippi. He left ten sailors and a friar on the frigate and the rest — thirty-four men, three friars, and the women and children — in the fort.

When La Salle's party returned to the fort in March 1686, they did not know where they had been, except that it was northeast from Matagorda Bay. They had reached a large river which they thought might be the Mississippi, except it flowed east. The built a redoubt by the river — probably the Colorado — and left some men there. It was never found and the men were never heard from again. By then the frigate had sunk in a storm and several men had drowned.

A couple, expecting a child, asked La Salle to marry them when he returned to the fort. He refused, citing their unequal social status. He said the child was entitled to a title of nobility. The problem was solved when the baby died shortly after birth.

La Salle was sick with a recurring fever. For three weeks he was barely conscious. Then he recovered, and, in April, set out again to search for the Mississippi. This time he took twenty men. They followed a route further inland, crossing the Colorado and Brazos Rivers, and reaching a point northwest of present Houston.

They described passing through a beautiful country with delightful fields and prairies bordered by vines, fruits, and groves, and filled with many herds of wild cattle. They met many Indians, some mounted and riding in Spanish saddles. The Indians were friendly — probably Caddoes — and they invited the French into their villages, some as large as three hundred cabins.

The Indians told of a cruel, wicked, white nation to the west. La Salle told them his nation was at war with that nation and the Indians asked the French to stay and go to war against the Spaniards. La Salle begged off but promised to return with "numerous troops."

By late May, only nine men remained with La Salle. Five had become sick and got permission to return to the fort. Five deserted, and one was lost while hunting. La Salle had narrowly escaped drowning while crossing the Brazos. He again fell ill. They waited two months for his recovery and returned to fort, defeated again in their search for the Mississippi. On the return journey, one of the men was carried off by a crocodile, probably in the Colorado River.

They reached the fort in late August. The people were in despair, regarding the country as "an abode of weariness and a perpetual prison." In early January, 1687, La Salle tried once more to find the Mississippi. Before he left he addressed the people left behind. Confident this time he would reach New France, he promised them he would return. His address at a midnight Mass on Twelfth Night was eloquent and brought tears to many eyes.

As his small party struggled around the shore of Lavaca Bay, they passed by a hundred or more rude crosses marking earlier graves. No one knew for sure how many others had died, their bones stripped by vultures and scattered by wild animals. Thirty one were left at the fort. Most died that summer. The rest were raided by Karankawas that fall. Four children were carried off by Indian women; the rest butchered.

La Salle and the sixteen men with him had learned from the previous attempts the best trail to the north. Every morning the men made moccasins from fresh-killed buffalo. As the footwear dried during the day, it shrunk to fit, but they had to soak their feet in water to remove the moccasins in the evening. Sometimes they would stop a few days to hunt and re-provision. Then the men would build a small stockade for protection from Indians, and the priest would set up an altar for Mass.

On March 16 near present Navasota, Texas, eight men were sent out on a party to hunt buffalo. That night after a heated argument, one of the men, backed up by two others, killed three men of the party as they slept.

Two days later, La Salle, with the priest and an Indian guide, set out to learn why the hunting party had not returned. When they reached a stream, La Salle saw Jean l'Archevique, one of the hunting party, on the opposite bank. He called out but received a rude reply. Just as La Salle rebuked the man, two shots were fired from a clump of reeds. One missed; the other hit La Salle in the temple, killing him instantly.

The grave of the 43-year-old explorer, if there ever was one, has never been found. Some reports say his body was stripped, dragged into the brush, and left to the mercy of wild animals. Further arguments among the survivors resulted in disputes about which way to travel. Some wanted to continue north to New France; others to return to Fort St. Louis. These disputes resulted in more shootings and deaths. A few reached New France that fall, and eventually Montreal and France.

In March, 1689, on the Pecos River in west Texas, Juan Xaviata, principal chief of the Cibolo and Jumano nations, handed some papers to Juan Retana, a general in New Spain stationed at the Presidio de Conchas. Xaviata had received the papers from an Indian messenger who said he had been asked by a Frenchman to forward them on to the Spanish. One of the papers, a picture of a French sailing vessel, bore this handwritten note in the margin: "Sir: I do not know what sort of people you are. We are French. We are among the savages. We would like very much to be among Christians such as we are. We know well that you are Spaniards. We do not know whether you will attack us. Gentlemen if you are willing to take us away, you have only to send a message, as we have but little or nothing to do. As soon as we see the note, we will deliver ourselves up to you." The note was signed, Jean l'Archeveque.

The note was unnoticed for two and a half centuries until discovered by a researcher in a musty archive in Spain. Its pitiful plea for Christian companionship from one of the conspirators in the murder of La Salle shows they would prefer imprisonment from the people they set out to conquer to their present life. It is an interesting footnote to the French invasion of New Spain.

Suggested reading: John U. Terrell, *La Salle, the Life and Times of an Explorer* (New York: Weybright & Talley, 1968).

FRONTIER DEFENSE

Over two and a half centuries of northward exploration by the Spanish preceded the westward movement that too often is used to define the course of Old West history. The farthest the Spanish came north was in 1720 in present Nebraska. They were protecting their frontier against the French, but they were stopped by Pawnee Indians.

Fears of French invasion from the east were heightened in 1686 when La Salle landed on the Texas coast. He had already descended the Mississippi, claimed its drainage area for France, and named it after his king, Louis XIV. Later he asserted that the Mississippi and the Rio Bravo (Rio Grande) were the same river, and the Spanish worried about this challenge to their sovereignty.

The governor of New Mexico, Antonio Valverde, appointed Pedro de Villasur to lead an expedition in 1720 to see if the French were encroaching into Spanish territory. Valverde, a soldier from Castile, had led an expedition the previous year into what is now eastern Colorado, almost a century before Zebulon Pike, coming from the east, saw the peak that was named for him.

But in 1720 Valverde chose to stay in Santa Fe and let Villasur, also a soldier from Castile, lead the expedition of forty soldiers, sixty Indians, three settlers, a priest, a French interpreter, and a scout. They rode out of Sante Fe on June 16.

By August 6, when they crossed the Rio Jesús María (South Platte), they hadn't seen a sign of a Frenchman. They decided to look for the Pawnees, well-known at that time for being friendly to the French, and see if they knew of any French in the area.

Three days later they crossed the San Lorenzo (North Platte) just above where it merged with the Jesús María. They called the combined river the Rio Chato and followed it downstream to a point between the present towns of Maxwell and Brady, Nebraska, where they stopped. They stopped very near the later location of Fort McPherson, a key point for westward emigrants as they traveled more than a century later over the Oregon Trail.

A large village of Pawnees were living on an island in the river. About two dozen came to the water's edge to talk peace. Villasur was pleased when they requested an interpreter, and he sent Juan de l'Archeveque, across the stream with the customary presents. (The Frenchman had been in La Salle's expedition and helped murder its leader. Generations later his descendant, Sostenes l'Archeveque, would be the most hated killer in New Mexico and West Texas.)

The interpreter shouted across the river that the Indians said they knew nothing of any French, and they would not let him come back. Villasur wanted to cross the river immediately and force out the desired information. His officers, impressed that Pawnees had just captured some of the Spaniards' Indians bathing in the stream, urged caution. After some discussion, the officers prevailed and the expedition crossed to the south side of the river and went into camp.

At daybreak on August 13, when the Spaniards were roping their horses and preparing their packs, the Pawnees and some French (according to the reports of survivors) attacked. Villasur was killed in front of his tent before he had time to get his weapon. Also killed were five corporals, nineteen soldiers, a settler, the interpreter, the priest, the scout, and eleven Indians with the Spanish. No one knows how many Pawnees were killed, but the Indians did not pursue the survivors. Apaches, friends of the Spanish, cared for the survivors as they straggled back to Santa Fe.

The Pawnee attack was a crushing blow for the New Mexican settlements. They had lost a third of their best soldiers. Valverde blamed "heretical Huguenots" (French Protestants) for the disaster and asked for thirty soldiers to replace those lost.

When Valverde reported to the Viceroy in New Spain, he said Villasur's attackers included two hundred Frenchmen. There is apparently no information from other sources that any French were within hundreds of miles of the attack. Valverde also reported that the Apaches had offered to ally with the Spanish if a new expedition to avenge the killing was mounted against the Pawnees and the French.

After an investigation in the City of Mexico, Valverde was fined two hundred pesos (fifty for masses for the souls of the fallen soldiers and one hundred fifty for the purchase of religious objects and ornaments for missions). The investigators found him guilty of selecting an incompetent officer to lead the expedition.

This little-known battle almost four hundred years ago was a sidelight to the European war between France and Spain. It highlights a common misconception about the directionality of Old West history. Perhaps it also says something about commonalities over those centuries in international relations, variations in military skill, and propaganda.

Suggested reading: Alfred B. Thomas (ed.) *After Coronado* (Norman: University of Oklahoma Press, 1966).

THE FRENCH IN THE NORTHERN PLAINS

Pierre Gaultier de Varennes was born in Three Rivers, New France, in 1685. He was the youngest of nine children, and his father was governor of that important colony on the St. Lawrence River, just as his mother's father had been earlier.

Pierre joined the army when he was twelve. After fighting in New England and Newfoundland, he went to France at age eighteen to serve against the English in the War of the Spanish Succession. Two months before his twentieth birthday he received nine wounds in the Battle of Malplaquet. Left for dead on the battlefield, he was a prisoner of war for fifteen months and then honored by promotion to lieutenant. But King Louis XIV was broke and couldn't pay the salaries of new officers, so Pierre returned to New France, accepted a lower commission as ensign, and opened a trading post near Three Rivers.

Pierre took the name, La Vérendrye, the one by which he is commonly known. He married in 1712 and fathered four sons in five years, all of whom took part in his explorations.

Northern explorers had long believed the Pacific Ocean could be reached by going up the Missouri River, crossing some mountains, and descending a westward-flowing river without great difficulty. Almost a hundred years before Lewis and Clark, Vérendrye proposed to cross the continent by building a chain of fur posts, each post furnishing the funds and manpower to build the next in line. Vérendrye did not get very far, but no one had any idea how far it was across the continent. George Washington would not begin surveying in western Virginia until decades later.

Without the drive or color of Radisson or Mackenzie, Vérendrye overcame obstacles with patience and determination. Using a crude bark map obtained from an Indian, he persuaded the governor of New France to commit a hundred men and supplies for exploration to the west. Then he persuaded Montreal merchants to advance trade goods on credit. In 1731 the middle-aged soldier and trader left Montreal with three of his sons, Jean-Baptiste, 18, Pierre, 17, and François, 16, along with his nephew, Christophe Dufrot, and a hard-looking crew of hunters and packers. They went through Michilimackinac and across Lake Superior to the Grand Portage. There, Vérendrye split the expedition, sending part on ahead under his nephew to build Fort St. Pierre at the mouth of Rainy Lake. A year later, Vérendrye built Fort St. Charles on the west shore of Lake of the Woods.

At Fort St. Charles, Vérendrye stopped a tribal fight between the Sioux and the Crees. He ordered them, in the name of the king, to stop fighting, promising them guns, axes, and gunpowder if they obeyed.

In the 1733-34 winter a delegation of Crees and Assiniboines from Lake Winnipeg visited Vérendrye at Fort St. Charles. They told about a tribe of light-skinned Indians, some with light or red hair, who lived in the distant southwest. These Indians — the Mandans — lived in earthen houses inside palisaded forts, practiced agriculture, domesticated animals, and had iron knives. They lived by a large river, and they had tunnels connecting their forts to the river. Learning that the French were exploring to the west, these Indians had asked the Assiniboines to tell the French that they looked forward to meeting them. They lived in a large territory, without a mountain. Some of it was prairie land; the rest tall timber. Vérendrye had heard about the strange tribe before and was determined to meet them.

Vérendrye sent Jean-Baptiste back with the Crees and Assiniboines, and this oldest son built Fort Maurepas at Lake Winnipeg later in 1734.

The only food Vérendrye brought out from the settlements was corn and coffee, along with garden seeds. The explorers relied on game, fish, and wild oats, from which they made bread. Most of their supplies had been sent ahead with Jean-Baptiste, and the supply canoes, expected in Fall 1735, never arrived.

The winter of 1735-36 was so severe the men at Fort St. Charles were living on roots, shoe leather, and their dogs. Then sons Jean-Baptiste and Pierre arrived in June from Fort Maurepas with news that Vérendrye's nephew had died on the trail. Vérendrye decided to send Jean-Baptiste with a party of nineteen *voyageurs* and the missionary, Father Aulneau, back to Michilimackinac to hurry up the supply canoes.

Times had been hard for Father Aulneau. The ship that brought him from France was plague-stricken. The trip to Fort St. Charles had been difficult for him and he had just weathered the year of famine. Now the young priest was going back to Michilimackinac so he could attend a religious retreat.

The twenty-one man party left Fort St. Charles in three canoes on June 5, 1736. Vérendrye knew a war party of Sioux were in the vicinity, probably hunting their Cree enemies, so he urged his son to be particularly vigilant.

Two supply canoes arrived at the fort on June 17. The men had seen and heard nothing of Jean-Baptiste's party. Vérendrye was glad to see his youngest son, Louis-Joseph, who came up with the supply canoes, but he worried about the oldest.

Two days later, Vérendrye sent nine men downstream to search. That party returned with the terrible news. The local Crees had always been friendly, so Jean-Baptiste had not posted sentries when they camped on an

23

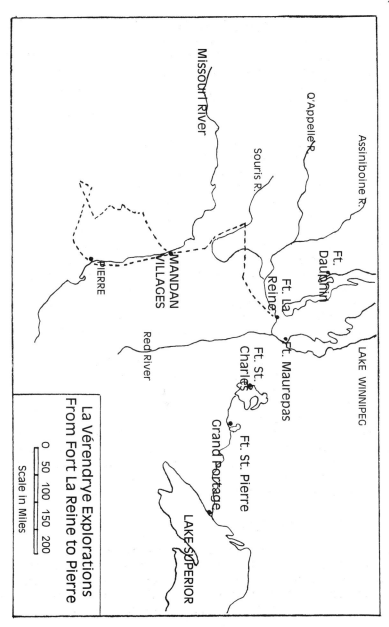

La Vérendrye Explorations
From Fort La Reine to Pierre

Scale in Miles
0 50 100 150 200

island their first night. Soon asleep in anticipation of an early morning start, no one heard the seventeen Sioux when they slipped out of their canoes to look at the sleeping men. Just as silently they slipped back in their canoes and carried the news to the rest of their marauding band.

Something had happened at Fort St. Charles which Vérendrye had no knowledge of at the time. Happy with some new firearms and probably warmed with some old brandy, a few mischievous Crees had fired from the fort at some Sioux wandering nearby.

"Who - Fire - On - Us?" the outrageous Sioux demanded.

"The French!" laughed the Crees.

The wandering Sioux were part of a band of one hundred thirty who learned of the sleeping *voyageurs.* By the time the Sioux reached the island, the Frenchmen were paddling away in the ghostly gray of dawn. The Sioux followed. When the fog rose at sunup, Jean-Baptiste landed his crews for breakfast.

The carnage apparently took little time. At a place ever since called Massacre Island, the bodies of the twenty-one men, each beheaded, were found in the sand. Two of the heads, Jean-Baptiste's and the missionary's, were never found, probably kept as trophies of war. It appeared the Sioux had killed them all without warning. Later on a Sioux admitted the massacre but claimed they did not intend to kill the missionary. He said Father Aulneau was struck suddenly by a hot-heated warrior, seeking distinction for bravery. There is also a report that as soon as the slaughter was over, a loud clap of thunder was heard, and the Indians fled in fear, thinking the Great Spirit was angry about the killing of the first martyr in the West.

Eight hundred furious Crees begged Vérendrye to let them avenge the murders. The veteran of Malplaquet exhorted them to not go to war.

Now, with his nephew and oldest son dead, Vérendrye had to decide whether to give up or continue exploring. He went to Montreal, seeking more financial help to hold off his creditors. Then, in 1737, he went to Fort Maurepas for a council with the Assiniboines and Crees. They offered to guide him to the River of the West, which was thought to flow to the Pacific. He declined, saying it was too late in the season, but he left gifts for the Assiniboines to take to the Mandans, together with the message that the French wanted to establish contact with that tribe and would build a fort near the junction of the Red and Assiniboine Rivers to be closer to them.

Vérendrye returned to Quebec and came back west the next year. Youngest son Louis-Joseph, 21, was put in command of Fort St. Charles, and the other two traveled on with their father. After stopping at Fort Maurepas, they traveled southwest and built Fort La Reine.

On October 18, 1738, Vérendrye, with twenty-four men including his sons plus two servants and twenty-five Indians, set out on foot to search for the Mandans. They reached the tribe on November 28, and were surprised that they looked no different than the Assiniboines. They had received the gifts that the Assiniboines had passed on. Vérendrye was told the tribe had six forts and the one they had reached was the only one not on the big river. Shortly after reaching the fort, Vérendrye's new bag of gifts was stolen. One of the Assiniboine guides, on leaving to return to their homeland, scolded the Mandans for the theft. The Mandan chief replied that they had searched everywhere for the bag, and he wasn't convinced that it wasn't stolen by the Assiniboines.

About a week after the explorers arrived, the Assiniboines were still in the Mandan village, and the Mandans were upset about how much they were eating. The Mandans started a rumor of an expected Sioux attack, and the Assiniboines headed for home.

Unfortunately the interpreter the explorers had hired had fallen in love with an Assiniboine woman on the trip down, and she refused to stay with the Mandans. So the interpreter ran off with her, probably joining the homeward-bound Assiniboines a day or two later. The interpreter was a Cree, but he could speak Assiniboine, as could a few Mandans. One of the Vérendrye sons was fluent in Cree, but no one could speak Mandan. Now the explorers were reduced to sign language.

The Mandans had heard of white men from the south who had horses, and who were impervious to injury while mounted because they were covered in iron. Vérendrye wanted to learn more, but the communication problem forced his return to Fort La Reine. He left two Frenchmen behind to learn the Mandan language.

Vérendrye was sick and they didn't get back to the fort until February 10, 1739. It was a difficult winter trip over the wind-swept northern plains, where buffalo sought shelter in wooded ravines and foxes and coyotes burrowed deeply in the snow.

"I have never been so wretched from illness and fatigue in all my life as on that journey," Vérendrye reported to his governor.

Two years passed before Pierre Vérendrye (the son) and one of his brothers returned to explore in summer, 1741. In October Vérendrye returned to Fort La Reine after a two year absence. By then Pierre (the son) had built two new trading posts in their continuing competition with the Hudson's Bay Company. Fort Dauphin was on Lake Winnipegosis, and Fort Bourbon was on Cedar Lake.

In 1742 François and Louis-Joseph started an exploration that would take them farther into the northern plains than any other Frenchmen had gone. They left Fort La Reine on April 29, reaching the Mandan Villages on May 19. On July 23, with the two Frenchmen who had been left there in 1739 and two Mandan guides, they set out — this time mounted — to the southwest, hoping to find the way to the Western Sea. Exactly where they traveled is unclear because the account, written by François after the journey, is sometimes confused in its directions and dates.

The best bet is that they traveled west into the North Dakota Badlands where they lighted fires on White Butte, hoping to find some people. They had seen none for twenty days. About mid-September they saw smoke and found a village of Blackfeet or Arapahoes. By then both their Mandan Guides had been released to return home. They reached a village of Kiowas about the middle of October. About a month later they were among the Pawnees, a strong, well-mounted people. Their principal chief knew a few words in Spanish and told of a battle in which they had destroyed Spanish who were searching for the Missouri River. (Just twenty-two years before, the Pawnees had routed the Pedro de Villasur expedition from Santa Fe, which was reconnoitering the northeast frontier of Spanish territory.)

After leaving the Pawnees they reached the Arikaras. By then (March 1743) they had given up the idea of reaching the Western Sea or the Spanish settlements, and they headed north toward the Mandan villages. While at the main fort of the Arikaras, on March 30 they secretly buried a lead tablet on a hill. This tablet was found by high school students on February 17, 1913, on a hill just across the Missouri River from Pierre, South Dakota.

After passing a village of "Sioux of the Prairies" (either Yanktonai Sioux or Cheyennes) they reached the Mandans on May 18. They were back at Fort La Reine by July 2.

During their year-long exploration, the youngest Vérendrye brothers, following their father's work, discovered information that contributed greatly to the success of Lewis and Clark over sixty years later. It also tells us much about the distribution of Indian tribes on the northern plains before the Teton Sioux left their eastern woodlands.

On December 6, 1749, the greatest French explorer in North America, again making plans for more western exploration, died suddenly in Montreal.

Suggested reading: G. Hubert Smith, *The Explorations of the La Vérendryes in the Northern Plains, 1738-43* (Lincoln: Univ. Of Neb. Press, 1980).

INDEPENDENCE DAY IN THE SOUTHWEST

The year 1776, when thirteen of England's North American colonies declared independence, was a momentous year on the eastern part of the continent. But just as important in United States history were two journeys that year in the distant Southwest. One of them was originally scheduled to start July 4, the day the bells rang in Philadelphia.

Spanish officials in the new world wanted ties between their California province, founded a century before, and their older New Mexico province, but no one knew the way between them. Three exploring Franciscan fathers, one a Daniel Boone in clerical garb, discovered the path.

In October, 1775, Father Francisco Garcés left his mission at San Xavier del Bac, south of present Tucson, to accompany an expedition led by Captain Juan de Anza, laying out an overland route between the missions in Mexico and those in California. The year before he had ridden on an earlier Anza expedition all the way to Mission San Gabriel. This time he would leave Anza at the Colorado River.

Father Garcés left the expedition at the rancheria of Yuma Indian Chief Palma on January 3, 1776. After ministering to the Indians, he left on February 14 to begin a long, lonely horseback ride.

He rode up the Colorado River to the Mohave villages near present Needles. There he turned west to Soda Lake and discovered the Mohave River. He followed that river and crossed over Cajon Pass into the San Gabriel Valley. When he reached the mission at San Gabriel he continued northwest, crossing Tejon Pass into the San Joaquin Valley, going north as far as present Fresno.

From there he looked east and thought he saw a possible pass through the Sierra Nevadas. He returned to the Colorado River by way of a pass he had discovered — since called Tehachapi. There he picked up a Mohave guide, and continued his journey toward the Hopi villages in northeastern Arizona.

Garcés crossed the Black Range and continued along the present location of the Santa Fe Railroad to Peach Springs. Following a trail that no white man had seen for over two hundred years, he continued northeast across Havasupai Canyon, visiting the Indians who lived there. He saw the deep gorge of the Grand Canyon at Quetzal Point, backed off, and continued east, crossing the Little Colorado to reach Oraibi, the main Hopi pueblo, on July 2.

Near midnight on July 3, 1776 — the bells would have been ringing in the east — Garcés wrote a letter about his journey, mentioning the gap he thought he saw in the Sierras. He hired an Indian boy to carry the letter to the Zuñi Indian villages, where he thought another friar might be stationed. When Garcés returned to his mission in September, 1776, he had been gone eleven months and had contacted about twenty-five thousand Indians. To this indomitable Franciscan, spreading the gospel required sweat and muscle as much as preaching.

In 1775, a few months before Garcés started his lonely ride, the provincial governor of New Mexico asked Father Silvestre Vélez Escalante, based at Zuñi, for a report about the hostile Hopis and the chances of finding a way through their land to Monterey, the capital of the Province of California. Escalante rode to Oraibi, the same Hopi pueblo where Garcés would later write his letter. There Escalante found a Havasupai Indian who had just come from the west to trade.

Escalante had a long smoke and a fruitful talk with the Havasupai, who drew a charcoal map on an equestrian breastplate to show the way through his land to the Colorado River, far to the west. Escalante carried that information to the governor in Santa Fe, little dreaming that it would lead to his own Homeric journey the next year.

In 1776 Father Francisco Atanasio Domínguez was sent to Santa Fe to take charge of all the Franciscan missions in the province of New Mexico. Thus, the expedition that was ordered to start on July 4 to look for a route to California which would avoid both the hostile Hopis in what is now northeastern Arizona and hostile Apaches further south is called the Domínguez-Escalante expedition. However, Father Escalante was the primary explorer. His diary of their journey contains details of flora, fauna, terrain, and climate, observations about favorable places for settlements, and thorough cataloging of the Indian tribes they met. It is one of the greatest source documents in American history.

Trouble with Comanches and an illness of Father Escalante delayed the start of the ten-man expedition until July 29. Garces' letter had reached Santa Fe by then, whetting the explorers' appetites to find a way from Santa Fe to Monterey.

The explorers followed an old trappers' trail — it was probably an Indian slave dealers' trail before that — northwest up the Chama River to cross the continental divide. One of the party got a good dunking in crossing the Chama when his horse went completely under in a sink hole. When they crossed the canyon on Amargo Creek, about a mile above its junction with the Navajo River, they called it the Canyon of Deceit on "account of a certain incident." They don't tell us what the incident was.

They forded the San Juan and camped on August 8 on the bank of the Las Animas about four miles south of present Durango, Colorado. They crossed the La Plata and Mancos Rivers and followed the Dolores to near present Egnar. Turning northeast, they camped near the intersection of

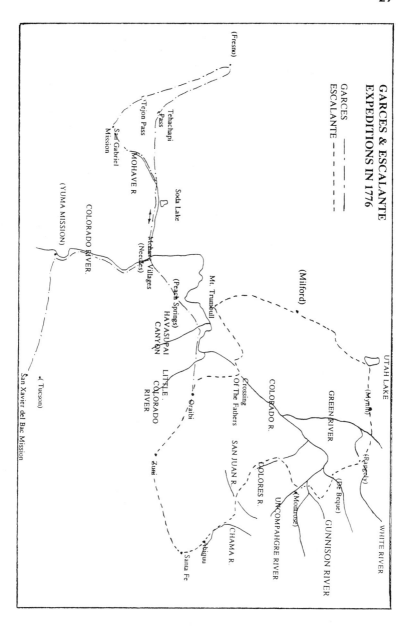

GARCES & ESCALANTE
EXPEDITIONS IN 1776

GARCES ————·——
ESCALANTE — — — — —

Highways 141 and 90, just west of Naturita. Then they crossed the Uncompahgre Plateau to present Montrose.

They camped on August 26 near the present Ute Tribal Museum where Highway 550 crosses the Uncompahgre River. Their next campsite was about two miles north of Olathe on the east side of the Uncompahgre.

On August 28 they camped on the south bank of the Gunnison River. A day and a half layover was needed to convince the local Ute Indians to furnish them guides. Following up the North Fork of the Gunnison on September 1, the Utes begged them to turn back, saying the Comanches on ahead would cause trouble.

The friars talked to the Indians about the Christian faith, and said it was wrong to have multiple wives. They had told their companions in the beginning to not bring trade goods lest the Indians think their purpose was more lowly than spreading the faith. On this day they were chagrined to discover that the interpreter and his brother had sneaked articles along, and were secretly trading with the Indians. They deplored the lack of faith of these men, saying, "how very unfit they were for ventures of this kind."

They continued up the north fork of the Gunnison and Muddy Creek, turning northwest to cross the Colorado River on September 5 near present De Beque. Continuing northwest through an area of Indian pictographs, they camped on September 9 at the intersection of the White River and Douglas Creek, just north of present Rangely.

They crossed the Green River — the largest one they had come to, which they called the San Buenaventura — on September 13, about six miles north of present Jensen, Utah.

They laid over for two days to rest the horses. One apparently did not need much rest. One of their Ute guides jumped on it and the horse raced away. It tripped and fell, throwing the "horse-breaker a long distance off." The Fathers were afraid their guide was seriously hurt, "But God was pleased that all the damage was borne by the horse, which got its neck completely broken."

On September 16 they saw signs that made them fear they were being scouted by Indians. They didn't know if they were Utes or Comanches. The guide said they were Comanches, scouting Utes.

They swung around to the west, crossing the Uintah River and following the Duchesne River to camp at present Myton, Utah, on September 17. That day they saw the ruins of a very ancient pueblo, which apparently has not been found since.

They continued west, near present Highway 40 and under Strawberry Reservoir, reaching Utah Lake on September 23. After visiting the Ute villages and admiring the scenery for two days, they headed southwest along the present route of Interstate 15. One of their Ute guides deserted them on October 5. That night they had a heavy snowfall, which continued all the next day. They were in the Beaver River bottoms, about twenty miles north of present Milford, Utah.

Unable to travel at all on October 7 for the cold and snow, they made less than ten miles on the eighth. Now they were drawing near the latitude of Monterey, their goal. They were low on provisions and afraid their last two guides would also desert. To the west they saw only formidable heights covered with snow — no low altitude pass, mentioned by Father Garcés (there is none). Domíngues and Escalante decided to turn back to Santa Fe. They continued south, hoping to cross the Colorado River and turn southeast toward Zuñi.

But others, looking forward to developing trade with California, complained about giving up their goal. A few days before, Don Bernardo Miera said they had not progressed very far west since leaving Santa Fe. Now he claimed they could be in Monterey in a week. Finally, on October 11, the fathers, upset that "the business of earth was being sought first and foremost over the business of Heaven," agreed to cast lots and let God decide. They had used this method before in deciding which of two trails to take when they were in doubt.

Domíngues said Don Bernardo Miera would be the leader if the lot fell to Monterey. Then he asked them all "to subject themselves to God's will and, by laying aside every sort of passion, beg Him with firm hope and lively faith to make it known to us." They prayed the rosary and recited psalms with the Litany, and cast the lots. The return to Santa Fe won.

They continued south, crossing present Highway 15 at Ash Creek, just north of Pintura. Their last two guides deserted on October 13. Two days later they crossed the Virgin River and camped just across the Arizona border.

Near Mt. Trumbull they began looking for a place to cross the Colorado River. They had to turn back northeast. On October 26 they tried to cross near the mouth of the Paria River. Their two best swimmers bundled their clothing on their heads and made it across, the clothing being washed away. The swimmers returned, apparently naked for the rest of the journey.

The next day they made a raft, but the poles couldn't reach bottom and the wind blew them back. They also feared for their horses, as both banks had quicksand. By this time they were eating their horses and what piñon nuts they could gather.

On November 6, following a heavy hail and thunderstorm which stopped after they implored relief by reciting the Virgin's Litany, they found a fordable place. In eleven days since their attempted crossing at the Paria River, they had traveled about twenty miles. But they crossed the next day.

They had to cut steps in the stone cliffs with their axes before the horses could get down to the river. The mile-wide ford was called the Crossing of the Fathers until it was buried by Lake Powell.

They continued south and east. On November 13 they caught a porcupine and feasted on the "flesh of the richest flavor." But it only whetted their appetites, so they killed another horse. They reached Oraibi on November 16 and Zuñi eight days later. They enjoyed a long rest at Zuñi, leaving December 13 and reaching Santa Fe on January 2, 1877, after a journey of more than five months and two thousand miles.

Although Domínguez and Escalante had more men, including an engineer with an astrolabe to determine latitudes, their journey was just as impressive as the lonely trip of Garcés. They explored more unknown land than either Daniel Boone had or Lewis and Clark would.

In the same year the English colonists were celebrating their newly-declared freedom, the Spanish colonists were learning that a long-desired land route existed between the old settlements in the Rio Grande Valley and the newer ones on the California coast. Spain's two colonizing thrusts north from Mexico could now be tied together.

Father Garcés, after his long ride, returned to the Yuma tribe on the lower Colorado River. A mission was established there in 1780. On July 17, 1781, while Garcés was saying Mass, the mission was attacked. Chief Palma asked his warriors to spare the missionaries, but two days later their bodies were found clubbed to death. The carnage resulted in the killing of forty settlers, twenty children, and thirty-one soldiers.

Father Escalante remained in New Mexico for several years, serving various pueblos. In 1780, on his way to Mexico City for medical care, he died in Parral.

Father Domíguez was recalled to Mexico to answer charges made by disgruntled brethren whom he had disciplined during his tour of inspection. He spent the remaining thirty years of his life in various missionary assignments, trying in vain for vindication of his conduct and recognition for a lifetime of service to the Church.

The 1776 explorations of these Franciscan Fathers left no question that New Mexico, Arizona, California, Nevada, Utah, and Colorado would stay intact under Spanish Rule. That vast area did stay intact until 1848 when it became part of the United States.

The year 1776 was as important in the southwest as it was in the northeast.

Suggested reading: Herbert E. Bolton, *Pageant in the Wilderness* (Salt Lake City: Utah Historical Society, 1950).

PETER POND

S ome fur traders were interested in exploration only as it increased their trading areas. Others were primarily explorers, and interested in fur trading as a way to finance the satisfaction of their insatiable thirst to learn what was over the next mountain range, down the next river, or across the next sea. Peter Pond was one of these. While he never realized the fame of Alexander Mackenzie, the first man to cross North America, he was important in Canadian exploration. In fact, Mackenzie was his protégé and learned well from a master.

Pond began his journal: "I was born in Milford in the countey of New Haven in Conn the 18 day of Jany 1740 and lived there under the Government and the protection of my parans til the year 56. Beaing then sixteen years of age I gave my parans to understand that I had a strong desire to be a Solge. . . . But they forbid me, and no wonder as my father had a larg and young famerly. I just begun to be of sum youse to him in his afairs. Still the same Inklanation & Sperit that my Ancestors Profest run thero my Vanes. It is well known that from fifth generation downward we ware all Waryers Ither by Sea or Land and in Dead so strong was the Propensatey for the arme that I could not with stand its Temtations."

The 16-year-old boy left his father's shoe repair shop and joined the army. He won a field commission as captain in the French and Indian War which, as the prime cause of the Seven Years War in Europe, ended in 1763 with the Treaty of Paris transferring French Canada to the British. At war's end Pond entered the Mississippi fur trade, but he left for the Canadian Northwest in 1775 after he killed a man in a duel. As Pond described it: "We met the next morning eairley and discharged pistels in which the pore fellowe was unfortenat."

Pond wintered at Fort Dauphin on Dauphin Lake in 1775-1776. In the spring he set out for the Saskatchewan River, passing La Verendrye's Fort Bourbon on Cedar Lake, which had fallen into decay. He wintered at the Forks of the Saskatchewan for the next two winters as he traded with Indians.

In Spring 1778 several fur traders gathered at Cumberland Lake and turned their trade goods over to Pond. He set out with four canoes for the Athabaska Country to establish a trading post in the best place he could find. He went up the Churchill River to Ile à la Crosse Lake which was the farthest point to the northwest yet reached by an European. From there he would be exploring what had only been known by reports of Indians.

Pond continued northwest by Lake Clear, Buffalo Lake, and Lake La Loche. A thirteen-mile difficult portage over the height of land dividing the

Hudson Bay drainage from the Arctic Ocean drainage brought them to the Clearwater River, a tributary of the Athabaska.

The portage, usually called Methye Portage, became famous in later years, both as the gateway to the immense water systems of the far northwest and as one of the most beautiful scenic vistas on the continent. Alexander Mackenzie and all the others who followed Pond praised the beautiful valley at the foot of the portage. As Mackenzie described it: "The valley is about three miles wide, confined by two lofty ridges, displaying a most beautiful intermixture of wood and lawn, and stretching on till the blue mist obscures the prospect." The ridges were covered by stately forests, broken up by green fields in which buffalo and elk were feeding.

For the first time in northwestern exploration a westward flowing river had been found. Pond knew from the Indians the river did not go to the much sought after Western Sea, but he hoped that following it he could find one that did. After eight miles down the Clearwater, his canoes swept out into the Athabaska and they turned north for thirty miles, where he unloaded and built a rude fort for a trading post.

The post was Pond's headquarters for the next six years as he explored from the Saskatchewan to Lake Athabaska and west up the Peace River. Called Old Fort Pond, the post was the only one in that part of the country until 1785. It was the father to Fort Chipewyan, for many years the most important establishment of the North West Company.

Pond was a morose, unsociable man, seeing offense when none was intended. He never found happiness among his own race, but preferred the wilderness and its more savage inhabitants, with whom he had much in common. After an argument between Pond and a trader, Jean Etienne Wadin, and Wadin's clerk at Lake La Ronge, Wadin was shot during the night. He died, and Pond and the clerk were tried for murder. Their acquittal on a technicality did not end the suspicion that they were responsible. Pond was also implicated in the death of another trader, John Ross.

Pond told his government about the Russians establishing a trading port on the North Pacific, and he warned that the newly formed United States would have easy access to the northwest through the posts the French had given up at the termination of the French and Indian war. He urged that the North West Company be encouraged in the prosecution of their discoveries so "a firm footing may be established as will preserve that valuable trade from falling into the hands of other powers."

Peter Pond certainly did his part.

Suggested reading: Lawrence J. Burpee, *The Search for the Western Sea* (Toronto: Musson Book Company, 1908).

A WOMAN'S VIEW

Louise Boyd explored the arctic, but that area has never been part of the old west. Wives accompanied soldiers and settlers in early Spanish exploration, but they didn't write about it. The wife of an English sea captain is the only one to write about a pre-Lewis and Clark exploration. Her name was Frances Hornby Barkley. Just married, she sailed with Captain Charles W. Barkley in 1787 to the northwest coast of North America. Captain Barkley was a poacher, sailing an English ship with an English crew under an Austrian flag to conceal his illegal activities, but this had no effect on his wife's observations. He returned on a second voyage in 1792 , and that is when her diary provides a fascinating glimpse of an early exploration from a woman's point of view.

They reached Yakutat Bay off the Gulf of Alaska on August 18, 1792, with Mt. St. Elias and Mt. Fairweather in view. Her diary continues:

"They appeared as most disgusting objects covered with dirty sea-otter skins, with fur to the skin, the leather tanned red and filthy beyond description. It was here we saw women with those pieces of shaped wooden lip ornaments, which are described in Captain Cook's voyages — if such a frightening appendage can be called ornamental, a thing that distorts the mouth and gives the whole features a new and most unpleasant character. The piece of wood is inserted into a slit made in the under lip when the females are about 14 years old, and it is replaced from year to year, larger and larger, until in middle age it is as large as the bowl of a table spoon and nearly the same shape, being concave on the inside of the lip, which it presses out from the gum, thereby showing the whole of the teeth and gums — a frightful sight at best, but still worse when the teeth are black and dirty, which was invariably the case, also generally uneven and decayed. This odious mouthpiece so completely disfigured them, that it was impossible to tell what they would have been without it, for even their complexions could not be ascertained, their skins being besmeared with soot and red ochre.

"Their hair is dark and shiny and appeared to be kept in good order, parted in the middle and kept smooth on each side behind the ears and tied behind the top in a knot. The men, on the contrary, have their hair matted and daubed with oil and ochre. The dresses of both sexes are made with the skins of animals, sometimes with the fur on and sometimes without. The women seldom wear any valuable furs; the men sometimes wear sea-otter skin of which they well know the value, and will strip themselves whenever they can make a good bargain."

After a violent gale made them stand out to sea, they sailed south to Sitka Sound, anchoring in a cove. The next day the natives showed up in canoes, their numbers increasing each day. "They got so bold and troublesome at last that it became difficult to avoid disputes, they stealing every article that they could lay their hands upon, stripping them when they went on shore, and upon the slightest offense presenting their firearms at us, the use of which they perfectly knew, but we conjectured had never felt the effect of, and certainly not of our great guns. Capt. Barkley, on one or two occasions, had our great guns fired off to astonish them, but they only seemed to think him in play. Thank God, we left them in ignorance of their deadly effect, but as they saw the trees shivered and broken by the cannon shot, they must have been aware of what mischief they could do.

"Once in particular, Capt. Barkley saw several war canoes with his night glass stealing along under the shadow of the land on a fine moonlight night, and as we were very indifferently manned, he was suspicious of their intentions. We therefore had the whole broad-side fired off over their heads, which made a tremendous noise among the trees. Every canoe scuttled off, but we kept perfect silence on board that they might not think we were alarmed. Early next morning they came alongside again, dressed in their war dresses and singing their war songs and keeping time with their paddles. When they had paddled three times around the vessel they set up a great shout, they pulled off their masks, resumed their usual habits, and exhibited their sea-otter skins for trade, giving us to understand that they had been on a war expedition and had taken these skins from their enemies. They never alluded to the firing but went on trading as if nothing had passed. They are a very savage race, and their women are still more frightful than the women of (Yakutat) Bay, the disgusting mouthpiece being still larger than theirs. In fact, the mouthpieces of the old women were so large that the lip could not support them, so that they were obliged to hold it up with their hands and to close their mouths with great effort. When shut, the under lip entirely hid the upper one and reached up to the nose. This gave them a most extraordinary appearance, but when they opened it to eat , no description can be given of what it is like, for they are obliged to support the lip whilst they opened their mouths, and then they throw the food into their mouths, throwing back their heads with a jerk to prevent the food lodging in the artificial lip or saucer, which is concave, and when let down receives whatever escapes the right channel. How any rational creature could invent such an inconvenient machine I am at a loss even to guess, as their is no stage of it that has the most distant appearance of ornament even in young women."

Suggested reading: "The Diaries of Frances Hornby Barkley" in Donald A Barclay (ed.) *Into the Wilderness Dream* (Salt Lake City: University of Utah Press, 1994).

BOLD VISIONS

Although John Ledyard barely touched the west coast of North America, his vision and plans for its exploration and his global adventures leave one breathless.

Fatherless at ten, the Groton, Connecticut, boy wanted to be a missionary to the Indians. In 1772, when he was twenty-one, John entered Dartmouth College, newly moved to New Hampshire from Connecticut, where it had been called Moor's Indian Charity School.

John liked the Indians; once he left college to live with some, but his adventurous spirit chafed at the discipline of college life. Headstrong, emotional, athletic, and charming, he persuaded friends to help chop down a tree and hollow it into a fifty-foot dugout. He sailed it down the Connecticut River, and became a sailor in New York on a ship carrying mules to Africa.

Back in London in 1775, he enlisted as a Marine with James Cook's third voyage, which reached the west coast of North America in 1778. Traveling along Vancouver Island, Corporal Ledyard saw metallic goods of European origin that must have followed Indian trade routes from Hudson Bay. If a white man's goods can cross America, he wondered, why couldn't a white man?

The question persisted as they sailed up an arm of the Gulf of Alaska — now called Cook's Inlet — and threaded their way through the Aleutians and across Bering Strait until turned back by Arctic ice. Cook sailed back to the Sandwich (Hawaiian) Islands, which he had discovered earlier on that voyage. He went ashore, guarded by his Marines, but the natives killed him, along with four of the Marines. The other Marines, including Corporal Ledyard, fought their way clear and swam to safety.

Back in England in 1780, Ledyard refused to serve against his countrymen in the Revolutionary War. However he was shipped to New York on a transport after the war had shifted to the south. He deserted and fled to Connecticut for employment in his uncle's law office. There he wrote his own account of the expedition of England's greatest explorer. The publication in 1783 of his *A Journal of Captain Cook's Last Voyage to the Pacific* beat the British Admiralty's official publication by a year.

Still fascinated by the idea that a white man might cross North America, he almost talked Robert Morris of Philadelphia, the bankroller of the American Revolution, into financing an expedition. But the country was in a depression and Morris was tired. For a time Ledyard had John Paul Jones interested, but that, too, collapsed.

Ledyard's war-weary Connecticut neighbors probably thought the

Appalachian Mountains — or at most the Mississippi River — bounded the new nation. But Ledyard had seen the Pacific shore and the man-made articles there, and he continued with his dream. He went to Paris to see the new Minister to France, Thomas Jefferson.

Ledyard had met Benjamin Franklin, the aging previous minister, and Jefferson thought at first he had just inherited Ledyard. But when they met they became friends, and Ledyard ignited the same fire in Jefferson's mind that burned in his own.

Ledyard told Jefferson about seamen trading trinkets to the coastal Indians for sea otter skins, which they then traded in China for oriental goods, that sold in England for great profits. He was the first American to see the possibilities of this triangular trade. By then Jefferson had read Ledyard's book and considered the man a genius.

Ledyard told Jefferson of his failure to promote an expedition by sea out of America. Jefferson suggested he travel overland across Europe, Russia, and Siberia, cross the Pacific in a Russian trading ship, drop down to the latitude of the Missouri River, and return to the east coast by foot. Ledyard was enthusiastic, and Jefferson obtained permission from Catherine the Great, Empress of Russia, for Ledyard to cross her empire.

Ledyard set out from London in early 1787, accompanied only by two dogs. Traveling by foot and hitching rides by boat, he reached St. Petersburg in June. Continuing by foot and hitching rides in Russian three-horse carriages, he reached Irkutsk in August, Yakutsk in September. But that was a close as he would get to his destination, Okhotsk. Catherine the Great had changed her mind. She had him arrested and, after wintering in Irkutsk, he was brought back to Russia and deported across the border into Poland.

But it's hard to keep an explorer like Ledyard down. He returned to Paris and London, and found an organization in England that wanted to know more about the mysterious Niger River in Africa. After a last visit to Jefferson in Paris, still convinced that someday he would find a way to cross North America, he sailed to Cairo in January, 1879. He wrote two letters to Jefferson from Cairo. There he died under mysterious circumstances on January 10. He was thirty-eight, and he had come a long way from Groton, Connecticut.

Suggested reading: Donald Jackson, *Thomas Jefferson & the Stony Mountains* (Urbana: University of Illinois Press, 1981).

A GOOD MAN IN THE MOUNTAINS

George Drouillard, half-Indian hunter and scout, was the prize recruit for the Lewis and Clark expedition. He was expert in sign language, and the leaders offered him a lieutenant's pay of twenty-five dollars a month to sign up as interpreter. Not until December, 1803, after Drouillard had sized up the two captains and their men in their first winter camp on the Mississippi River, across from the mouth of the Missouri, did he agree to go.

Drouillard was a skilled hunter and a brave man. He was excused from guard duty to be the principal hunter for the group. He was often away for a week at a time, as he scoured the country for deer, bear, elk, and buffalo. Once, he shot a charging grizzly through the heart with a single-shot muzzle loader. The bear fell dead at his feet.

Drouillard's ability to find game and keep the explorers on the best trail was invaluable. One time Lewis left a note for Clark, telling him which trail to follow. He put the note on a fresh-cut stick, and apparently a beaver carried it away. Clark never saw the note. He led his group up the wrong river. Drouillard found them and brought them to the right trail.

The scouting and sign language skills of Drouillard were particularly important in August, 1805, as the expedition struggled up the headwaters of the Missouri. They had cached their large boats at the great falls, and they were moving ahead in canoes. But they had to get horses from the Shoshones in order to cross the mountains. They worried about meeting the Shoshones, a warlike tribe, who might resent the invasion of their homeland. The first contact would be crucial, and the explorers were quite sure Indian scouts were watching as they struggled through the Three Forks country.

To complicate matters, Captain Clark had seriously injured his foot, and had to travel slowly. So Drouillard, Captain Lewis, Charbonneau — another interpreter — and a soldier went ahead, up the Jefferson River. They watched carefully, hoping for a peaceful contact with the Indians.

The advance party finally met the Shoshones on August 13, after crossing Lemhi Pass. Drouillard's sign language made the first contact peaceful. The chief embraced Lewis affectionately and rubbed his cheek against the captain's.

The Shoshone Indians were out of meat. Lewis wrote that he viewed "the poor, starved divils with pity and affection." The Indians gave horses to Drouillard and the soldier so they could hunt. Most of the young warriors followed as Drouillard chased, but could not catch a herd of antelopes.

Two days later Lewis persuaded the Indians to provide more horses and travel with him as he went back to find Clark. The extra horses would be necessary to carry up the baggage.

Then first morning going back, Drouillard killed a deer. The famished Indians devoured the deer without waiting to cook it. They even ate the soft parts of the hooves. By the end of the day, Drouillard had killed two more deer, and all the Indians ate their fill.

On the next day, August 17, occurred one of the most remarkable coincidences in the history of the West. Lewis sent Drouillard, accompanied by one Indian down the mountain to look for Clark. The Indian soon returned, reporting with excitement that he had seen the other whites. Until then, in spite of Lewis's assurance and Drouillard's furnishing them with meat, the Indians suspected that Lewis was leading them into a trap where they would be attacked by their enemies.

Sacagawea, wife of Charbonneau, was a Shoshone. She had been captured some years before and sold to the Missouri River tribe, who lived where Lewis and Clark had spent the previous winter. The explorers were not surprised when Sacagawea saw the Shoshones and recognized her own tribe. But they were as amazed as she and the Shoshones when she recognized the chief as her own brother, Cameahwait.

The next day was Lewis's thirty-first birthday. He expressed in his journal his regrets that he had done little for the betterment of the world and a resolve that he would try to do better. As one of the leaders in the most important exploration ever made by this country, Lewis may never again have doubted his own significance. If the grateful Shoshones had not provided horses, the expedition would probably have failed; at least, Lewis thought so. The importance of Drouillard should not be underestimated.

A river was named for Droiuiiard. The stream, in present Idaho, is now called the Palouse.

Four years after the Lewis and Clark expedition, while he was trapping on the Jefferson River, the stream he had followed to the dramatic meeting with the Shoshones, Drouillard was killed by Blackfeet Indians.

Suggested reading: Bernard De Voto (ed), *The Journals of Lewis and Clark* (Boston: Houghton Mifflin Co., 1953).

NEW GENES IN THE MOUNTAINS

Old York and Rose were slaves, owned all their lives by Virginians John and Ann Clark. Their son, Ben, and the Clarks' son, William, grew up to be lifelong friends. When William Clark, then an army officer, was assigned to help lead an exploring expedition to the Pacific Ocean, he immediately thought of his old friend. The roster of fifty-one men in the Lewis and Clark Expedition included Ben York.

York was a huge man with herculean strength. He had a sense of humor to match. His own companions enjoyed him, as did all the Indians he met. York was also a gentle, caring man when those qualities were needed. The expedition only lost one man, Sergeant Charles Floyd. He died of a ruptured appendix near what is now Sioux City, Iowa. The faithful York nursed Floyd during his short illness.

The Arikaras, whom the expedition met in present South Dakota, had never seen a black man. They were astonished with York. They examined him from head to toe. The giant's jet-black, kinky hair was a marvel to them. York told the Indians he had been a wild animal, living on the flesh of children, until he was caught by Clark and tamed. He told the Indians that he found children very good to eat, but his twinkling eyes and deep laughter convinced them he would not devour their babies.

Arikaras traditionally showed hospitality by turning their women over to guests. One of the Indians invited York into his lodge and presented him to his wife. Then the warrior retired outside the lodge to wait. Soon one of York's companions came looking for him. In the words of one of the party's diarists: "The gallant husband would permit no interruption until a reasonable time had elapsed." Clark noted in his journal that the Arikara women were fond of caressing his huge, black servant.

The expedition spent the 1804-1805 winter with the Mandans in present North Dakota. This tribe showed its hospitality like the Arikaras. No one in the expedition had to be lonely that winter. The Mandan women thought York was the best looking man they had ever seen, and they fought for the chance to bear his children.

On December 8, Clark's journal mentioned that sudden cold weather had caught a hunting party off guard. Several of the men suffered from frostbite. "My servant's feet were also frosted and his penis a little," Clark wrote.

In late June near the great falls of the Missouri, York just missed one of the most exciting adventures on the expedition. He and Clark, along with Sacagawea, her baby and husband, had gone on ahead of the rest. A heavy hailstorm drove Clark and the Indians to cover in a deep ravine. A flash flood would have drowned Sacawagea and her family but for Clark's alertness and skill. The raging torrent was waist deep on Clark as he pushed the others to safety. But York missed the rescue as he had gone ahead to hunt buffalo. When the storm came up, he returned to look for the others. He was relieved to see them climbing out of the ravine.

York was also much admired by the Shoshones when that tribe was reached. The explorers carried merchandise to trade for horses, but the Shoshones were more interested in York than in the trade goods.

Charles M. Russell made thirteen paintings and sketches based on the Lewis and Clark Expedition. The most famous is a large mural in the Montana House of Representatives. It shows the explorers meeting the Flathead Indians in early September. The painting shows the excitement of the Indians as they pointed to the tall giant, towering above his horse. Perhaps they had heard about him from the tribes farther east. Russell also painted the interior of a Mandan lodge, showing York as the center of attention.

When the explorers visited the Mandan village on their way downstream in 1806, they saw many dark-skinned babies with kinky hair. It appeared that the cold snap in December, 1804, had even enhanced the expansion of the growing gene pool of the upper Missouri River tribes. Several persons reported that York's progeny could be traced among the Mandans and other tribes until the 1890s.

York's day in the sun ended in September, 1806, when the explorers returned to the East after the greatest exploration in our nation's history. Clark left the group in Louisville, Kentucky, and York went with his master.

Five years later Clark freed all his slaves, but he remained in close contact with York for the rest of York's life.

Suggested reading: Charles G. Clarke, *The Men of the Lewis and Clark Expedition* (Glendale: Arthur H. Clark Co., 1970).

LED ON BY THE LOVE OF NOVELTY

For Jedediah Smith exploring was incidental to the fur trade. He was an amateur compared to two professionals, Lewis and Clark. Yet he saw much more of what became the western United States than they. During his short, eight-year career he was the first to go overland to California, and first to travel the length of California and Oregon by land.

Smith knew the country from the Missouri River to the Pacific, from Canada to Mexico. He survived the three worst disasters of the fur trade only to die a lonely death on the Santa Fe Trail.

In the rough, unruly west where it was said God kept on his own side of the Missouri, Smith was an unlikely explorer. Mild and modest, quiet and gentle, he did not smoke, chew, or swear. He took a little wine or brandy only on formal occasions, and he always carried his bible in his shirt pocket.

But no one was tougher. Hardship and suffering never held him back. His courage under fire, intelligence, and ability to command were outstanding. He went west in 1822 with only his rifle, his bible, and the clothes he wore. When he returned eight years later, he was recognized as the best of the mountain men.

In Spring 1822, aged twenty-three, Smith showed up in St. Louis. His ancestors, descended from old American stock in Massachusetts, had migrated west through New York and Pennsylvania to Ohio. The young man saw William Ashley's famous help wanted ad for a hundred enterprising young men to ascend the Missouri, and he left in May on Ashley's second boat, hired as a hunter.

After a winter of hunting and trapping on the Yellowstone, Smith was asked by Major Andrew Henry to carry a message downstream to his partner, Ashley. Returning upriver with Ashley in 1823, Smith was in the famous fight with the Arikaras, the worst disaster in the Western Fur Trade. Thirteen men, a sixth of Ashley's force, were killed in the attack at Grand River on June 1, with nine more wounded.

Ashley retreated downriver. Smith volunteered to continue upriver, through the heart of Arikara country with a message for Major Henry. With one companion, he reached Fort Henry, riding part of the way on horses stolen from Indians.

Henry left twenty men behind to garrison the fort. The rest, including Smith, embarked down stream to rescue Ashley. The Arikaras waved happily from the river banks at these downstream travelers.

They reached Ashley early in July. By this time he was supported by regular army troops. As the combined forces continued upriver, the Ashley-Henry men were divided into two companies. Jedediah Smith was made captain of one of them. With the help of the army and some Sioux allies, the trappers got past the Arikara village with little trouble.

Henry traveled west to the Yellowstone. Ashley dropped back down river and dispatched Smith and a party of eleven west into the mountains of present Wyoming. Most of Smith's men had been with him in the worst of the Arikara attack. They included William L. Sublette, James Clyman, Thomas Fitzpatrick, and Edward Rose, all destined for fame in the fur trade.

On this journey, Smith was attacked by a grizzly bear. In the mauling, the bear's jaws crunched down on his head from his left eye to his right ear, scraping the skull bare, almost ripping off his ear. Without losing consciousness, Smith directed James Clyman in his surgical repairs. To the end of his life he wore his hair long to cover the horrible scars.

It was also on this journey, in March, 1824, that Smith's party discovered South Pass, the route over which thousands of emigrants would cross the mountains in later years.

Jedediah Smith was back in St. Louis with Ashley in October, 1825. Three weeks later he headed back west as Ashley's partner, leading seventy men, one hundred sixty horses and mules, and carrying trade goods worth twenty thousand dollars.

Nine months later Ashley sold out to a new firm, Smith, Jackson and Sublette. Smith was the senior partner. David E. Jackson was the discoverer of Jackson's Hole, and Sublette had been traveling with Smith since 1823.

In August, 1826, Smith led seventeen men southwest from the Great Salt Lake in search of the legendary Buenaventura River which was supposed to flow through the western mountain range to the Pacific. They reached Mission San Gabriel, near present Los Angeles, in November.

Jedediah Smith was impressed with the mission. It owned a herd of twenty thousand cattle which supported the hide and tallow trade along the Pacific Coast. Smith sent a message to Governor Echeandia in San Diego, requesting permission to travel on to San Francisco Bay. After weeks of hemming and hawing, the suspicious governor finally gave him permission to leave the province the way he had come. He could not proceed northwest toward the Russian establishment on Bodega Bay.

Smith recrossed the San Bernardino Mountains and turned north along the edge of the Mohave Desert to enter the long San Joaquin Valley. He did not believe he was violating the governor's order, as mountain men thought of California as the settled strip along the coast and the trail of missions. It did not include the Sierra Nevadas, where they would trap beaver and continue their search for the great, westward-flowing river.

In May, 1827, Smith reached the American River, where they tried to cross the mountains. By this time they had fifteen hundred pounds in beaver pelts, and the snow was too deep. Five horses died in the attempt. They turned back south seventy-five miles and started up the Stanislaus.

The snow was still too deep for the whole party, but Smith, leading two men, seven horses, and two mules, crossed at Ebbett's Pass. They lost only two horses and one mule. Eating their horses as the animals gave out, the three men reached Bear Lake for the 1827 rendezvous on July 3.

Ten days later the rendezvous ended, and Smith was back on the trail to Southern California with eighteen men. He expected to relieve the men he had left, but he revealed an explorer's motivation when he wrote, "I was also led on by the love of novelty common to all which is much increased by the pursuit of its gratification."

About the middle of August, when they were crossing the Colorado River, Mohaves attacked, killing ten of the party. Left with eight men, surrounded by four to five hundred hostile Indians, armed only with butcher knives and five rifles, their horses and most of their provisions stolen, Smith probably thought back four years to the Arikara attack, the only fur trade disaster to exceed in casualties the one he was then facing. But the situation now was beyond compare to the earlier one. There were no friendly Indians and no game in the vicinity, no troops down river.

They reached the river bank, moved to a cottonwood thicket, hacked small trees into a slight breast works, and made lances from poles and butcher knives. The men asked Smith if he thought they could defend themselves. He didn't think so, but he wanted to reassure them. "I told them I thought we would. But that was not my opinion. I directed that not more than three guns should be fired at a time and those only when the shot would be certain of killing."

As the Indians crept closer he finally told his two best marksmen they could fire. Two shots rang out, and three Indians fell, two killed and one wounded. Then the Indians "ran off like frightened sheep and we were released from the apprehension of immediate death."

They traveled in the desert all night, finding their first water at dawn. They had no way to carry water, and the August heat blistered, so they stayed at the spring until evening. Continuing in this manner, trying to follow a trail he had only traveled once and then in the daytime with horses, he finally got his men to the San Bernardino Valley. There they turned northwest, reaching the Stanislaus on September 18.

Smith stayed two days, got his men organized for trapping, and then took three with him to Mission San Jose. Not sure of what would happen, he told the men left behind if he did not return to go on to the Russian fort at Bodega.

The mission Fathers took the small party's horses and put Smith in the guard house. After more than three months of international complications involving Governor Echeandia, who was then in his office in Monterey, captains of English and other vessels in port at Monterey and San Francisco, confinements in San Jose and Monterey guard houses, a hearing before a military officer, and threats to transport Smith to Mexico at his own expense, Smith was allowed to lead his party north. A detail of Mexican soldiers led them, making sure the American spies had no contact with the Russians at Bodega. Smith turned his furs over to a vessel bound for the East, for a four thousand dollars.

Smith invested most of the fur money in horses, which would be worth much more at the 1828 rendezvous than they were in California. Counting the men from his 1826 crossing, he now had twenty men and over three hundred horses to lead.

In January, 1828, they trapped the lower tributaries of the San Joaquin — the Calaveras, the Mokelumne, the Cosumnes. They crossed the American River in late February. By late March they were at present Chico and could see Mt. Shasta in the far north. In April Smith had two close calls with bears. The first time he was afoot, and he dove into a creek to escape. The second time he was mounted, and the bear charged, grabbing his horse's tail. The horse dragged the bear fifty yards before it relinquished its hold.

In early April they crossed the Sacramento River just above present Red Bluff. Indians attacked as they headed into the Trinity Alps. The Indians followed them all the next day. Smith tried to be friendly, but the Indians' violent gestures and constant yelling persuaded him that force was necessary. The mild Christian man who had been through the two worst attacks in the fur trade finally gave the command to fire, and two Indians fell.

Moving three hundred half-wild horses and mules through the narrow passes, the vertical cliffs, the thick timber, and the tortured undergrowth of the Trinity Alps was difficult. Many days one or two would be crippled on the rocks or drowned in the river. Yet they had to keep on. And now in late April, they sometimes had to flounder through four-foot snow drifts.

By now the Indians they saw had long hair, dressed in deer skins, and wanted to trade for axes and knives. Smith knew they were approaching Hudson's Bay Company territory. As they came down the Trinity and the Klamath and struggled to the coast, they came to redwoods, the noblest trees Smith had ever seen. By late May they could see the ocean. By late June they had fought their way through the worst of the torturous ravines and dense undergrowth, had crossed the California River which still bears Smith's name, and crossed the Rogue River in Oregon.

As they neared the Coos River, the Indians became unfriendly, and their horses came into camp with arrows in their sides. In one three-day period they lost twenty-three animals, some to drowning, some to Indians. When they reached the Umpqua River on June 13, the Indians said they could go upstream about fifteen or twenty miles, cross a ridge, and reach the Willamette, which would take them to a Hudson's Bay post at Vancouver. The next morning, with two men and an Indian guide, Smith set out in a canoe to scout the way. He warned the men in camp to not, under any circumstances, let the Indians enter camp.

When Smith returned to the camp, it was strangely silent. An Indian on shore spoke to his guide, and the guide suddenly grabbed Smith's rifle and dove into the river. Hidden Indians fired on the boat as Smith paddled frantically to the opposite bank. He and his two men jumped out and ran to the ocean. The branch of the Umpqua on which he had been camped is still called the Smith River. The Indians had killed all but one man, who escaped by running away. The Umpqua Massacre, in which seventeen trappers were killed, was the third worst disaster to happen in the American fur trade. Jedediah Smith had survived them all.

Smith and his two companions went north to a Tillamook Village, where Indians guided him to Fort Vancouver. They reached the fort on August 10, two days after Arthur Black, the other survivor.

Smith returned to the Northern Rockies in March, 1829, and trapped the Blackfoot country with Jim Bridger. At the 1830 rendezvous he and his partners sold out to the five men who formed the Rocky Mountain Fur Company. Then he returned to St Louis.

He bought a house, but there was more exploring to be done, so he organized a caravan to Santa Fe. Leaving in April, 1831, he led seventy-four men and twenty-two wagons, half belonging to him. On May 27 the train had been out of water for three days while traveling between the Arkansas and the Cimmaron. Comanche Indians arriving at Santa Fe later in the year with some of Smith's property said he had ridden up to a water hole, where about twenty warriors were concealed, waiting for buffalo. In the fight Smith killed the leader of the band, but was himself killed.

Among the mountain men, he stood alone, and it was alone he died.

Suggested reading: Dale L. Morgan, *Jedediah Smith and the Opening of the West* (Lincoln: University of Nebraska Press, 1971).

BORN ON THE BORDER

Peter Skene Ogden's father was a New Jersey patriot when the Revolutionary War began. But the widower married a loyalist lady who influenced him to move to England and back to Canada, where he became a distinguished judge. There, in 1794, Peter was born, the youngest of their five children. The judge named his son after Peter Pond, a friend who had traveled widely in the North American wilderness. Peter's frequent saying that he was born on the border became a symbol of his long Canadian and United States exploration.

Peter read law for a time to please his father, but at age sixteen he began a seven-year employment with the North West Company at Fort Ile à la Crosse on the Beaver River west of Hudson Bay. Just a half mile away was a Hudson's Bay Company post. At that time the companies were in a bitter war over the fur trade in the western part of Rupert's Land.

In 1817 Ogden and his Cree wife crossed the continental divide and followed the Columbia River to Fort George (formerly Astoria) at its mouth on the Pacific. The fort had been sold to the North West Company four years before, during the war between England and the United States.

In 1820 Ogden became a partner in the company. By then he and his wife had two sons, Peter, three, and Charles, one.

In 1821 the North West Company merged with the Hudson's Bay Company (HBC). Leaving his wife and sons at Fort Ile à la Crosse in 1822, Ogden went to England to learn of the merged company's plans for him They made him a clerk of the first class with the salary of a chief trader. After visiting his aged parents, who had retired to England, he left for the northwest coast of North America. He reached York Factory on Hudson Bay in July, 1823. The company sent him to take charge of the Spokane District, the most important assignment it had ever given a man still in his twenties.

Ogden, leading two canoes with eight *voyageurs* left York Factory on July 18. When they reached Ile à la Crosse, Ogden learned that his sons were with their mother's family, and we have no evidence of what had happened to his Cree wife. The woman, whose name we do not know, is never mentioned again.

After a one day rest, Ogden continued west, turning upstream at the Athabaska. On October 3 they reached the Rocky Mountain House, the end of their canoe voyage. They changed to horses, and a week later were in valleys where the ice never thawed. They crossed the Rocky Mountain Divide at Athabaska Pass and took the Wood River down to the Columbia. On October 28, they reached Spokane House, headquarters of the district

Ogden was to command. It had taken three months and ten days from Hudson Bay.

Ogden met and married Julia, a Flathead woman, at his new post. About six years older than he, she lived with her tribe near the fort. Both her father and her husband had been killed by Blackfeet. Her mother had married François Rivet, who had been a member of the Lewis & Clark expedition and was now serving as interpreter for a Hudson's Bay expedition into the Snake River country of the United States.

At this time United States trappers, perhaps empowered by the notion of manifest destiny, were pushing west into the Pacific Northwest. The English, having already claimed Vancouver Island, were pushing just as hard to the south. An 1818 treaty between England and the United States allowed each country joint occupancy of the land between the Rocky Mountains and the Pacific Coast. The English decided to trap the Snake River region bare, leaving a fur wasteland to keep the Americans away. George Simpson, the new HBC Governor selected new Chief Trader Peter Skene Ogden to lead the expedition.

On December 20, Ogden led the expedition out of Flathead Post in present Montana. He had ten salaried trappers, forty-five free trappers (métis and eastern Indians who worked as independent contractors and paid the company for their supplies), thirty women and thirty-five children, including his own family. He told Julia he didn't see how she could possibly go with their two-month-old son. She replied that to an Indian the journey of life was made by men, women, and children together. If one part stopped while the other went on, they could never properly be together again. They stopped the day after Christmas while the wife of one of the free trappers had a baby. Ogden also had his father-in-law as one of his interpreters and two hundred sixty-eight horses.

One of the earlier HBC trappers in the Snake region was Finan McDonald. When he left he wrote: "I got home safe from the Snake Cuntre, thank God, and when that Cuntre will see me agane the beaver will have gould skin."

By the eighth day of the expedition, Ogden's trappers had caught three small beaver. His scorched earth policy was producing strange dividends.

On December 29 in bitter cold, Ogden rejoiced that only ten inches of snow was on the ground. That night Jedediah Smith and six of his men rode into their camp. They would stay with the expedition until they were out of dangerous Blackfeet territory. One might wonder if the American trappers knew the purpose of the English expedition. The next day they passed through Hellgate Canyon, near present Missoula, where Blackfeet and Piegan war parties had repeatedly attacked Flatheads.

They turned south up the Bitterroot Valley and over the continental divide into present Idaho. There one of the trapping parties was attacked by Blackfeet. Ogden complained that his free trappers spent too much time chasing buffalo. One day they killed thirty, but only brought three hundred pounds of meat into camp. The Indians said their horses were too tired from the chase — two of them had died — to bring the meat in. Yet the camp was very low on food.

On March 19, 1825, Jedediah Smith and his Americans left the expedition to cross the mountains on their own. Three days later a small party of trappers narrowly escaped from sixty Blackfeet. The attacks continued, and on April 8 the Blackfeet killed one of the free trappers.

They saw the American trappers from time to time, and the two trapping groups seemed to be playing cat and mouse with each other. In mid April Ogden reported that the Americans had been following them and were camped just three miles ahead. "It will avail them naught," Ogden wrote, "we have traps twelve miles ahead."

About that time they abandoned the Blackfoot River for the Portneuf. In one week in late April, they caught one hundred thirty-four beaver, which brought their total up to a hard-won thousand skins. In May, without knowing it, they crossed over the present north boundary of Utah into land claimed by Mexico. This diversion into previously unexplored country left Ogden's name on a river, a peak, and a city in Utah. His men also brought in the first reported sight of Great Salt Lake, which they said was as large as Lake Winnipeg.

The next day, May 23, saw the unfolding of an international drama. It started with the arrival of Etienne Provost, a free trapper of the Rocky Mountain Fur Company of St. Louis. With him were a free trapper who had deserted the HBC Snake River Expedition two years before, some Canadians, a Russian, and an old Spaniard. They seemed to be waiting.

In late afternoon another group arrived behind a waving American flag. Ogden recognized in it fourteen free trappers who had deserted his expedition. The newcomers joined Provost's group and they all rode into Ogden's camp, where their leader, Johnson Gardner, announced that since they all were in United States territory, everyone was free. He said the HBC trappers, whether company men or free trappers but indebted, could desert and join his group.

Ogden thought he was in joint occupation territory where the English and the Americans had equal rights of possession. But his over-charged and underpaid free trappers had no loyalty to the company, so he kept a strict watch to prevent their smuggling furs to the enemy camp. The next day Gardner ordered Ogden to return to Spokane House. Ogden said he would go when the English government told him to.

More free trappers deserted, saying the HBC men on the Columbia were villains but admitting that Ogden had treated them fairly. Ogden stood firm under the American guns until a free trapper took two company horses. Then Ogden grabbed the lead ropes and held on until the trapper paid for one horse and relinquished the other. Two free trappers paid their debts before leaving, but a dozen more escaped with their furs and without paying. Ogden, with only twenty trappers left of the original fifty-five, started back, still insisting he would not leave until the English government said he should.

Ogden and Gardner were mistaken about where the international incident took place. They were in Mexican territory, where both were trespassers.

On June 15, Ogden was back in Flathead country. A month later his second in command started back to Spokane House with two salaried trappers and twenty horses loaded with furs. There is no mention of Julia after this on the expedition, so she and the children probably went back at this time.

Ogden continued to trap with his reduced force, moving back south, then west and northwest until, on November 12, he reached Fort Nez Perces where the Snake empties into the Columbia. His total catch for the season was four thousand pelts. He had certainly done all he could to turn the Snake River country into a fur wasteland.

Ogden made a total of six trapping expeditions into what became the United States. He was the first to reach the Humboldt River and trace the route by which thousands of emigrants would come west. He trapped northern Utah and Nevada and explored the Pit River in California. In 1829-1830 he traveled from Fort Vancouver to the Gulf of California, going down through the Great Basin, and returning by the San Joaquin and Sacramento valleys. His explorations were almost as wide ranging as those of Jedediah Smith, with whom he often discussed the regions they explored, sometimes exploring together.

Following the 1847 massacre at the Whitman mission, Peter Ogden by great tact, decisive action, and high personal risk to himself, rescued the forty-seven survivors. Born on the border between England and the United States, this man of courage, intelligence, and integrity, who gave in to no one, served each country equally well. Born in Canada, he died in Oregon in 1854, aged sixty. His beloved Julia, surrounded by children, grandchildren, and great grandchildren, lived for thirty-two more years. She died in Canada aged ninety-eight and is buried there.

Suggested reading: Archie Binns, *Peter Skene Ogden: Fur Trader* (Portland: Binfords & Mort, 1967).

A DRY CROSSING

For two centuries Santa Fe was an island of settlement in a broad sea of unoccupied land. Then in the early 19th century, traders from the United States arrived over what would become the Santa Fe Trail. Soon after, Antonio Armijo made the first crossing from Santa Fe to California.

Armijo set out with sixty men from the outpost at Abiquiu on November 7, 1829. The beginning of his route was well known. Spaniards had long traded for slaves and furs with the Ute Indians to the northwest. Armijo knew that Jedediah Smith had traveled from Great Salt Lake to the San Gabriel Mission in Southern California three years before and repeated the journey the next year. Armijo hoped to use Smith's trail for the final part of his route. But first his pack train —- heavily loaded with trade goods — would have to find a way across the rough desert country along the present Utah-Arizona boundary.

They followed the traders' route to the four corners area. They came upon six Navajos when they crossed the San Juan River the second time. The explorers feared the Navajos, so when they reached a small Navajo village two days later on November 23, they traded eleven horses for a guide to lead them as far west as he knew the country. They hoped the Navajo guide would prevent other Navajos from attacking.

After passing north of Canyon de Chelly, they reached a canyon on November 30 which was so steep they had to unload the mules and carry the baggage up by hand. On December 5, after a day without water, they reached the mesa of the Rio Grande (present Colorado River). They forded the river at the same place used by the Dominguez-Escalante expedition in 1776. Armijo's men had to repair the 53-year-old stairs, cut into the steep sandstone walls by the earlier explorers, before their animals could climb out.

On December 10 the expedition found a Paiute village in the vicinity of the Paria River. Armijo, thankful for escaping trouble from the Navajos, described the Paiutes as a "gentle and cowardly nation." Three days later they reached a creek near present Fredonia, Arizona. By then they were melting snow for water.

The expedition intercepted Smith's trail when it reached the Severo (present Virgin) River, near Hurricane, Utah. Armijo knew that Smith had traveled south, down the east side of the Virgin, to cross the Colorado. Smith had then continued south to the Mohave villages and crossed the Colorado again to head west across the Mohave Desert to the mission at San Gabriel. Armijo wanted to cross the Virgin and proceed directly southwest.

The route would be shorter, and the two crossings of the Colorado would be avoided. On Christmas Day Armijo sent out a reconnaissance party to search for water along the proposed shortcut. Six days later the party returned. One man, Rafael Rivera, was missing. No water had been found, so the expedition continued south along the Virgin to its mouth on the Colorado. They continued west to the big bend in the river near present Las Vegas. Armijo was still wondering if they should try the short cut when Rivera, a seasoned explorer, walked into camp on January 7. He had found an Indian village on the Armargosa, a tiny stream that ran northwest into Death Valley.

Armijo led the expedition away from the Colorado, striking southwest across the Mojave Desert. They camped at the Little Spring of the Turtle (present Cottonwood Springs), and, after three more days — one without water — they reached the Amargosa. After more camps on small springs or in places where holes could be dug to water, and a few waterless days, they reached the Mojave River near present Barstow, California.

Armijo sent a party ahead to find a California rancho and bring back supplies. Then, as travelers had before and would later in the barren land, they sacrificed their horses for food. Armijo's journal for four days in January revealed these grim entries: "23 — we ate a horse; 24 — ditto; 25 — ditto; 26 — Ditto. We ate a mule belonging to Miguel Valdes."

The next day the advance party returned with supplies. The expedition reached Mission San Gabriel on January 31, after eighty six days of travel. The Californians were delighted to trade horses and mules, worth little to them, for the beautiful woolen goods the explorers had brought from the east.

Later travelers would move the center portion of Armijo's trail further north to avoid the parched land and deep canyons. To honor the original explorers in 1776, the traders called the route the Old Spanish Trail. The trail, somewhat south of the Domingues - Escalante route, is now recognized as a National Historic Trail.

Suggested reading: LeRoy R. Hafen, *Old Spanish Trail* (Glendale: Arthur H. Clark Co., 1954).

COVERT OPERATIONS

Benjamin Louis de Bonneville, French-born friend of LaFayette, graduated from West Point in 1815. After frontier service in the 7th Infantry, he took a two-year leave of absence in October, 1831, to organize, outfit, and lead a 110-man party into the West, whose principal mission is now obscure. Ostensibly it was a private venture by a dashing soldier of fortune, using family wealth, to study the possibility of making money in the fur trade. It appears he was engaged in a secret fact-finding mission to counter the British advance into the Northwest through its Hudson's Bay Company posts.

Bonneville did pay his own way, and his men trapped furs which he sold for his own account. But his relationship with President Andrew Jackson and the nature of his reports suggest a substantial government interest in the venture. Bonneville overstayed his leave by almost two years, but Jackson restored him to active duty. When he died in 1878, Bonneville was eighty-two, the oldest retired officer in the army.

The expedition left Fort Osage, near Independence, Missouri, on May 1, 1832, with Joseph Reddeford Walker as principal assistant. It traveled the route that would later become the Oregon Trail — up the Platte and Sweetwater Rivers and through South Pass. Bonneville built a post on the Green River, but it was so poorly located the trappers called it Bonneville's Folly and never used it. He established winter quarters on the North Fork of the Salmon among the Nez Perce and Flathead Indians. His hunters brought in reports of trouble with the Blackfeet Indians. Bonneville soon learned the Indians were completely under the influence of the Hudson's Bay Company.

The next year, after some differences with both the representatives of the American Fur Company and the free trappers known as the Rocky Mountain Men, Bonneville dispatched Walker and forty men to cross the Sierra Nevada into California. They went down the river discovered by Peter Ogden — later John C. Fremont would change the name from Ogden's River to the Humboldt — made a difficult crossing of the mountains in which they discovered Yosemite Valley, and wintered at Mission San Juan Bautista.

Shortly after Walker left, Bonneville wrote to General Alexander Macomb, Commanding General of the Army, and requested an extension of his leave of absence. His long letter summarized the information he had collected about the various Indian Tribes, the strength of the British outposts, the history of the British occupation, the extent of the fur trade, the quality of the timber lands, and the size of an American military force

Captain Benjamin Bonneville

Denver Public Library - Western History Dept.

large enough to take possession. He recommended that if possession was taken, it be done as soon as possible

He expected to establish his next winter quarters on the lower Columbia River, from where he would travel to California the next year. Perhaps he expected to meet Walker, but he didn't say so in his letter. He traveled extensively that summer through present Idaho, Utah, and Wyoming. The 1833-1834 winter quarters were on the Portneuf River, a tributary of the Snake, where Fort Hall would soon be built.

On Christmas day, 1833, Bonneville and three men set out to explore the Columbia River basin. They reached the Hudson's Bay post, Fort Walla Walla, in early March. The factor in charge of the post made it clear that his company was not looking for competition. Bonneville traveled down the river a short distance, but the Indians were all loyal to the Hudson's Bay Company.

Walker and his men returned from California in June, 1834, and rejoined Bonneville on the Bear River. Later that fall Bonneville returned to Fort Walla and was rebuffed again. Most of his exploration in 1834 was in Oregon and Idaho.

Bonneville set up winter quarters for 1834-1835 on the headwaters of the Bear River, near Great Salt Lake, not far from the headquarters of the previous winter. He returned to the United States in August, 1835. His leave of absence had long since expired — it had never been extended — and he had been dropped from army rolls. President Jackson restored him to active duty.

Bonneville reported that the country was much more extensive than he expected. He prepared a surprisingly accurate map to accompany his report. It shows the country from the Rocky Mountains to the Pacific and from the latitude of San Francisco Bay all the way to the forty-ninth parallel. It shows a Lake Bonneville, which we now call Great Salt Lake. However, Bonneville's name has been given to a large prehistoric lake that occupied most of present northern Nevada, northern Utah, and some of Idaho.

We do not have the contents of Bonneville's full report. But it would appear that covert operations are an old and respectable tool of governments.

Suggested reading: Washington Irving, *The Adventures of Captain Bonneville* (Norman: Univ. of Oklahoma Press, 1961).

A SHORT ROMANCE

Montgomery Pike Harrison, whose grandfathers were President William Henry Harrison and explorer Zebulon Montgomery Pike, graduated from West Point in 1847. Two years later he was chosen to lead infantry troops in a special exploration detail commanded by Captain Randolph Marcy. The detail also included a detachment of dragoons led by Lieutenant John Buford, who had graduated from West Point a year after Harrison. The two young officers enjoyed a good-natured rivalry.

Marcy headed west from Fort Smith toward Santa Fe, searching out a route across the southern plains for California-bound emigrants. In addition, the soldiers escorted those emigrants who were already on the trail that spring of 1849. The emigrants included former Army doctor John Conway, his wife, and their ten children. Seventeen-year-old Mary Conway was so beautiful that all who saw her were instantly infatuated. The expedition's physician was the first to propose. But the leading suitors were Harrison and Buford.

When the troops came to a large rock formation north of present Weatherford, Oklahoma, they named it Rock Mary. The enthusiastic soldiers thought the rock — two hundred feet around at the base and sixty feet high — was a suitable monument for the lovely girl.

When Mary had to make up her mind so the soldiers could get theirs back to exploring, she chose Harrison. Her parents were pleased with the engagement to the handsome, capable officer, but the impetuous young lieutenant wanted an immediate marriage. Doctor Conway told him he should wait until the family was settled in Los Angeles, where he planned to practice medicine.

"But we're leaving you at Santa Fe," the disappointed young officer said.

"I'm sorry, young man. Mary is so young, you know. Her mother and I want you to be patient. We're sure everything will work out."

The couple resigned themselves to the wait that seemed so interminable. For the rest of the journey to Santa Fe, they got together whenever Harrison's duties permitted. Whispered promises and discussion of future plans filled the precious moments.

After a tearful parting at Santa Fe, the emigrants went on, Harrison volunteered for a 600-mile expedition against the Apaches, and Marcy's men rested their mules.

When Harrison's detachment returned, Marcy continued his exploration. The detail marched south to Doña Aña and then turned east, looking for the shortest practical wagon road back to Fort Smith. They reached the Pecos River on September 16. They marched down the river and then northeast into the Staked Plains.

For nine days they saw little wood, less water, and no Indians. On October 3 they reached a beautiful spring, the location of present Big Spring, Texas. There they saw many signs of Comanches. They continued northeast, crossing small tributaries of the Concho and Colorado Rivers.

On October 7 Lieutenant Harrison rode away after the noon stop, saying he would examine a ravine about two miles away. When he did not return by evening, Marcy ordered the cannon fired, so his young officer, if lost, could hear the sound.

At daybreak the next day, Marcy sent another lieutenant and Black Beaver, the famed Delaware scout, out to look for Harrison, still missing. They found his tracks and reported that he seemed to have met two Indians. The three dismounted and had a smoke together in a grassy place. Then the Indians overpowered Harrison, and shot him with his own rifle. The scout found the body, stripped and scalped, among some rocks. The killers were never found. Black Beaver thought they were Kiowas.

The soldiers wept as they covered the body with tar and packed it into a sealed coffin for transportation back to Fort Smith. Marcy wrote: "A better young officer, or more courteous, amiable, and refined gentleman never lived. He was universally loved by all who knew him."

Montgomery Pike Harrison would never know that his friend and rival, John Buford, won recognition as the Union Army's best cavalry general in the Civil War. He would never know that his younger brother, Benjamin, followed their grandfather into the presidency. And he would never know that his beloved Mary became the wife of a New England sea captain shortly after she learned the tragic news.

Mary and her husband, whom she met in Los Angeles, had four sons and two daughters. She often told her children of her romance on the trail. She would have agreed with Marcy's description of the young officer she loved. He was the only casualty on the expedition.

Suggested reading: Grant Foreman, *Marcy & The Gold Seekers* (Norman: University of Oklahoma Press, 1939).

CROSSING THE SOUTH SASKATCHEWAN

On October 25, 1870, British Army officer William Francis Butler rode west from Fort Garry up the Assiniboine River. Butler had completed his term of service as intelligence officer for the Red River Expedition against the rebellious Metis. The new Dominion of Canada then asked him to explore further west toward the Northwest Territories and report on conditions prevailing among the Indians and traders of that almost uninhabited country. He was also asked to make recommendations about establishing and enforcing a system of law in advance of the influx of settlers that were expected as control of the land passed from the Hudson's Bay Company to the Dominion.

The Saskatchewan River Valley had been ravaged by smallpox, and the Canadian officials also sent vaccine with Butler. Carrying the vaccine in his Red River cart, Butler traveled with a Hudson's Bay Company official and a French-Indian helper.

Nights of sharp frost followed cold, sunny days as they moved northwest. The ice in the tributary streams seemed to reach out farther from the banks as they moved into higher and colder country. Far to the south they could see the Turtle Mountains on the border with North Dakota, and far to the north, dim against the horizon, they could see the tops of the Riding Mountains.

On October 30, they reached Fort Ellice, where the Assiniboine River turned north. As they forded the river, swift-moving masses of ice grated against the necks and shoulders of their horses. They continued west, toward the Saskatchewan.

Butler's French-Indian helper left him at Fort Ellice and was replaced by Daniel, an English-Indian half-breed. Butler only remained long enough at the fort to obtain warmer clothing. He replaced his boots and hat with fur moccasins and headgear. He rolled his stirrup irons with strips of buffalo skin. He also got large, fur mittens. Ahead of him lay three hundred miles without a fort or building of any kind, with treeless expanses sixty miles across.

On November 7, after a camp in below-zero weather, they reached the South Saskatchewan. Masses of ice reached hundreds of yards out into the river. In the center, bordered by ten-yard-wide banks of thin ice, a swift, black-looking current roared treacherously past.

They made a raft from a tarpaulin-covered wagon box. They launched it into a hole, chopped by Daniel, where the thick ice met the thin. Then Daniel got in and chopped a lane through the thin ice toward the open water. The tarpaulin leaked, and the raft had to be hauled out and emptied

WILLIAM F. BUTLER HUNTING BUFFALO ON THE SASKATCHEWAN PLAINS

Glenbow Archives, Calgary, Alberta

from time to time. By evening, Daniel had still not reached the open water.

Daniel cut through to the open water the next day, and Butler tried to make the first crossing to the other side. The current was too swift, and he failed, barely getting back without going under. Sharp ice had cut the tarpaulin. They replaced it with another sheet, and Daniel was able to get across, with the other two men pulling hard on long, buffalo-skin lines to keep the raft from disappearing downstream. By vigorous chopping, Daniel was able to get the raft far enough into the solid ice that he could step ashore. By then it was dark again, so they hauled the raft back and made camp for another night.

That night was bitterly cold, and the river froze completely over before morning. The men tested the new ice with their axes and sharp pointed poles. It was thin in places, but in others it rang hard and solid to their blows. They decided to try their lightest horse. They led the animal out with a long line tied to its neck. The ice sagged under the weight of the horse, but did not break, and the horse crossed safely.

Butler's saddle horse was next. Two of the men led the horse, and Butler followed close behind, ready to force the animal forward if necessary. When they reached the center of the river, the ice again bent downward. But, to Butler's horror, the ice suddenly broke and the horse plunged into a black, seething chasm. Butler, a few yards behind, leaped backwards to keep from going in himself.

The men watched, helplessly, as the horse floundered, trying to swim in its icy grave. Each time it reached the edge and tried to climb out, more ice broke away.

"Can we do nothing?" Butler screamed.

"None, whatever," they answered.

Butler ran back to the camp, picked up his rifle and returned. His hands trembled as he raised the weapon, aimed, and fired. The horse collapsed and disappeared under the cold, unpitying ice.

Butler returned to the camp and cried like a child. They did not try to cross any more horses. Butler and the Hudson's Bay man crossed on foot, caught the light horse, and walked on to Fort Carlton, twenty miles away. Daniel stayed behind with the rest of the horses, waiting for a few more nights of freezing weather.

Butler was pleased that his recommendations after the three thousand mile journey, mostly in the dead of winter, led to the formation of the Northwest Mounted Police, but he never forgot the saddle horse he lost at the crossing of the Saskatchewan.

Suggested reading: William Francis Butler, *The Great Lone Land* (Rutland Vermont: Charles E. Tuttle, 1968).

ORDERING INFORMATION

True Tales of the Old West
is projected for 38 volumes.

For Titles in Print,
Ask at your bookstore
or write:

PIONEER PRESS
P. O. Box 216
Carson City, NV 89702-0216
Voice Phone (775) 888-9867
FAX (775) 888-0908

Other titles in progress include:

Frontier Artists
Army Women
Western Duelists
Government Leaders
Early Lumbermen
Frontier Militiamen
Frontier Teachers

Ghosts & Mysteries
Californios
Doctors & Healers
Homesteaders
Old West Merchants
Scientists & Engineers
Visitors to the Frontier